Wind Tunnel Designs and Their Diverse Engineering Applications

Wind Tunnel Designs and Their Diverse Engineering Applications

Editor

Helmut Krakowski

scitus
academics

Wind Tunnel Designs and Their Diverse Engineering Applications
Edited by **Helmut Krakowski**

ISBN: 978-1-68117-664-2
Library of Congress Control Number: 2016946531

www.scitusacademics.com

Printed in United States of America on Acid Free Paper

Preface

Wind tunnels have evolved over the past thirty years as an essential aid to the civil engineering process. A wind tunnel is an aerodynamic tool used to learn the effects of air moving past solid objects. A wind tunnel consists of a tube passage mounted in the middle of the object under test. At the beginning of aeronautical research, the first wind tunnels were invented towards the end of the 19th century when many were trying to create successful heavier than air flying machines. The wind tunnel was conceived as a way to reverse the usual paradigm: instead of the air standing still and an object moving at speed through it, the same effect is obtained if the object was standing still and the air moved past it at speed. Wind tunnels are created for a specific purpose and speed range, and there are many types of wind tunnels and models. Applications for wind tunnel research range from regular airframe tests to basic research on the border layer, the slow moving air layer adjacent to anybody surface exposed to the wind. Measurements of other characteristics and air pressure at many points in the model provide information on the distribution of the entire wind load. Aerodynamic studies in wind tunnels, together with spacecraft and aircraft, are extremely profitable tools to solve design problems in automobiles, bridges, trains, boats and building structures.

This compendium is intended to provide invaluable insights into wind tunnel designs and their enormous research potential for students, industrialists, scientists, engineers, consultants and others.

Table of Contents

1

CHAPTER

Design Methodology for a Quick and Low-Cost Wind Tunnel

Miguel A. González Hernández[1], Ana I. Moreno López[1], Artur A. Jarzabek[1], José M. Perales Perales[1], Yuliang Wu[2] and Sun Xiaoxiao[2]

[1] Polytechnic University of Madrid, Spain

[2] Beijing Institute of Technology, China

1. INTRODUCTION

Wind tunnels are devices that enable researchers to study the flow over objects of interest, the forces acting on them and their interaction with the flow, which is nowadays playing an increasingly important role due to noise pollution. Since the very first day, wind tunnels have been used to verify aerodynamic theories and facilitate the design of aircrafts and, for a very long time, this has remained their main application. Nowadays, the aerodynamic research has expanded into other fields such as automotive industry, architecture, environment, education, etc., making low speed wind tunnel tests more important. Although the usefulness of CFD methods has improved over time, thousands of hours of wind tunnel tests (WTT) are still essential for the development of a new aircraft, wind turbine or any other design that involves complex interactions with the flow. Consequently, due to the growing interest of other branches of industry and science in low speed aerodynamics, and due to the persistent incapability of achieving accurate solutions with numerical codes, low speed wind tunnels (LSWT) are essential and irreplaceable during research and design.

A crucial characteristic of wind tunnels is the flow quality inside the test chamber and the overall performances. Three main criteria that are commonly used to define them are: maximum achievable speed, flow uniformity and turbulence level. Therefore, the design aim of a wind tunnel, in general, is to get a controlled flow in the test chamber, achieving the necessary flow performance and quality parameters.

In case of the aeronautical LSWTs, the requirements of those parameters are extremely strict, often substantially increasing the cost of facilities. But low turbulence and high uniformity in the flow are only necessary when, for example, laminar boundary layers have to be investigated. Another example of their use is aircraft engines combustion testing; this in turns requires a costly system that would purify the air in the tunnel to maintain the same air quality. Another increasingly important part of aircraft design is their noise footprint and usually the only way to test this phenomenon is in a wind tunnel.

In the automotive applications, it is obvious that the aerodynamic drag of the car is of paramount importance. Nevertheless, with the currently high level of control of this parameter and also due to imposed speed limitations, most of the efforts are directed to reduce the aerodynamic noise. The ground effect simulation is also very important, resulting in very sophisticated facilities to allow testing of both the ground effect simulation and noise production in the test section.

In architecture, due to the fact that buildings are placed on the ground and are usually of relatively low height, they are well within the atmospheric boundary layer. Therefore, the simulation of the equivalent boundary layer, in terms of average speed and turbulence level, becomes a challenging problem.

The design of the wind tunnels depends mainly on their final purpose. Apart from vertical wind tunnels and others used for specific tests (e.g. pressurised or cryogenic wind tunnels), most of the LSWTs can be categorised into two basic groups: open and closed circuit. They can be further divided into open and closed test section type.

For most applications, mainly for medium and large size wind tunnels, the typical configuration is the closed circuit and closed test chamber. Although, due to the conservation of kinetic energy of the airflow, these wind tunnels achieve the highest economic operation efficiency, they prove more difficult to design resulting from their general complexity. Hence, we will pay more attention to them in this chapter.

Apart from some early built wind tunnels for educational purposes at the UPM, since 1995 a number of LSWTs have been designed following the methodology which will be presented here. It focuses on the reduction of construction and operation costs, for a given performance and quality requirements.

The design procedure was first used for a theoretical design of a LSWT for the Spanish Consejo Superior de Deportes, which was to have a test section of 3,0 x 2,5 x 10,0 m^3 with a maximum operating speed of 40 m/s. Based on this design, a 1:8 scale model was built at UPM. This scaled wind tunnel has been used for research and educational purposes.

The second time it was during the design of a LSWT for the Instituto Tecnológico y de Energías Renovables de Tenerife (ITER). That wind tunnel is in use since February 2001, operating in two configurations: medium flow quality at maximum operating speed of 57 m/s, and high flow quality at maximum operating speed of 48 m/s. For more information visit www.iter.es.

Another example of this design procedure is a LSWT for the Universidad Tecnológica de Perú, which is now routinely used for teaching purposes. This wind tunnel is now in operation for about one and a half year.

At the moment the same procedure is being utilised to design a LSWT for the Beijing Institute of Technology (BIT). This wind tunnel will be used for educational and research purposes. It will have a high quality flow, up to 50 m/s, in a test section of 1,4 x 1,0 x 2,0 m^3. It will be used for typical aerodynamic tests and airfoil cascade tests (utilising the first corner of the wind tunnel circuit).

The design method to be presented in this chapter is based upon classical internal ducts design and analysis method, e.g. *Memento des pertes de charge: Coefficients de pertes de charge singulières et de pertes de charge par frottement,* I.E. Idel'cik [Eyrolles, 1986]. It also includes design assisting software such as a macro-aided Excel spreadsheet with all the complete formulation and dimensioning schemes for automatic recalculation. At the moment the best example of use of the method is the BIT-LSWT, mentioned above, as it has been defined using the latest and most reliable generation of wind tunnel design methodology.

2. MAIN DESIGN CRITERIA

The general layout of the proposed wind tunnel is shown in Figure 1. The airflow circulates in the direction indicated in the test chamber (counter clockwise in the figure). Upstream of the test chamber we find the other two main components of the wind tunnel: the contraction zone and the settling chamber. The other crucial component is of course the power plant. The remainder of the components just serve the purpose of closing the circuit while minimising the pressure loss. Nevertheless, diffuser 1 and corner 1 also have an important influence on the flow quality and they are responsible for more than 50% of the total pressure loss.

The design criteria are strongly linked with the specifications and requirements and those must be in accordance with the wind tunnel applications. The building and operation costs of a wind tunnel are highly related to the specifications and these are just a consequence of the expected applications.

In the case of the so called Industrial Aerodynamics or educational applications, the requirements related to flow quality may be relaxed, but for research and aeronautical applications the flow quality becomes very important, resulting in more expensive construction and higher operational costs.

The main specifications for a wind tunnel are the dimensions of the test section and the desired maximum operating speed. Together with this the flow quality, in terms of turbulence level and flow uniformity, must be specified in accordance with the applications. At this point it should also be defined whether all the components of the wind tunnel are going to be placed on the floor in a horizontal arrangement or in a vertical one, with only half of the circuit on the floor and the other half on top of it.

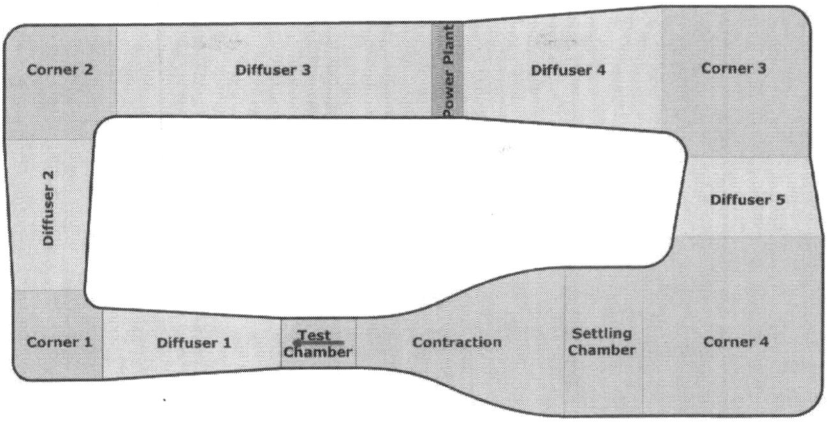

FIGURE 1. General layout of a closed circuit low speed wind tunnel. Figure labels indicate the part name, according to standards.

Flow quality, which is one of the main characteristics, is a result of the whole final design, and can only be verified during calibration tests. However, according to previous empirical knowledge, some rules can be followed to select adequate values of the variables that affect the associated quality parameters. The recommended values will be discussed in the sections corresponding to the Contraction, Settling Chamber, Diffusor 1 and Corner 1, which are the wind tunnel parts that have the greatest impact on the flow quality.

Once these specifications are given, it is very important to obtain on one side the overall wind tunnel dimensions to check their compatibility with the available room, and on the other side a preliminary estimation of the overall cost. The cost is mainly associated to the external shape of the wind tunnel and the power plant requirements.

For the benefit of new wind tunnel designers, a tool has been devised and implemented in an Excel spreadsheet (visit web page http://www.aero.upm.es/LSLCWT). Using this tool the designer will immediately get information about each part of the wind tunnel, the overall dimensions, the global and individual pressure loss coefficients, and the required power. This will be done according to the recommended input parameters and specification based on the intended use of the wind tunnel.

3. WIND TUNNEL COMPONENTS DEFINITION

In the following sections the design of each part will be thoroughly discussed and analysed in detail to get the best design addressing the general and particular requirements. Before dealing with each component, some general comments are given for the most important parts. In the case of the contraction zone, its design is crucial for achieving the required flow quality in the test section. In this sense, its contraction ratio, length and contour definition determine the level of uniformity in the velocity profile, as well as the necessary turbulence attenuation. It is crucial to avoid flow separation close to the walls of the contraction zone. At the stage of design, the most adequate method to verify that design meets those criteria is computational fluid dynamics (CFD).

Other important parts of the wind tunnel design worth mentioning here are the corners which incorporate turning vanes. Their aim is to reduce pressure loss and, in the case of the corner 1, possibly improve flow quality in the test section. The parameters to be considered in their design are the spacing between vanes (whether the space ought to be constant or not) and the possibility of expanding the flow (increasing the cross-section).

To complete the design process, the measurement equipment needs to be defined together with the complimentary calibration tests. Special attention needs to be devoted to the specification and selection of the balance for forces measurement, a device that is used to measure aerodynamic forces and moments on the model subjected to airflow in the test section. Since the drag force on test subjects can be very small and significant noise may be coming from the vibration of the tunnel components, such as the model stand, the true drag value may become obscured. The choice of an appropriate force balance is therefore crucial in obtaining reliable and accurate measurements.

The selection depends mainly on the nature of the tests. Wind tunnel balances can be categorized into internal and external ones. The former offers mobility since it is usually only temporarily mounted to the test section and may be used in different test sections. However, the latter has more potential in terms of data accuracy and reliability since it is tailored to a specific wind tunnel and its test section. Due to this reason, external force balances should be studied in greater depth.

3.1. Test Chamber

The test chamber size must be defined according to the wind tunnel main specifications, which also include the operating speed and desired flow quality. Test chamber size and operating speed determine the maximum size of the models and the maximum achievable Reynolds number.

The cross-section shape depends on the applications. In the case of civil or industrial applications, in most of the cases, a square cross-section is recommended. In this case, the test specimens are usually bluff bodies and their equivalent frontal area should not be higher than 10% of the test chamber cross-sectional area in order to avoid the need of making non-linear blockage corrections. Accurate methods for blockage corrections are presented in Maskell (1963).

Nevertheless, a rectangular shape is also recommended for aeronautical applications. In the case of three-dimensional tests, a typical width to height ratio is 4:3; however, for two-dimensional tests a 2:5 ratio is advised in order for the boundary layer thickness in the test section to be much smaller than the model span.

Taking into account that it is sometimes necessary to place additional equipment, e.g. measuring instruments, supports, etc., inside the test chamber, it is convenient to maintain the operation pressure inside it equal to the local environment pressure. To fulfil this condition, it is recommended to have a small opening, approximately 1,0% of the total length of the test chamber, at the entrance of the diffuser 1.

From the point of view of the pressure loss calculation, the test chamber will be considered as a constant section duct with standard finishing surfaces. Nevertheless, in some cases, the test chamber may have slightly divergent walls, in order to compensate for the boundary layer growth. This modification may avoid the need for tail flotation correction for aircraft model tests, although it would be strictly valid only for the design Reynolds number.

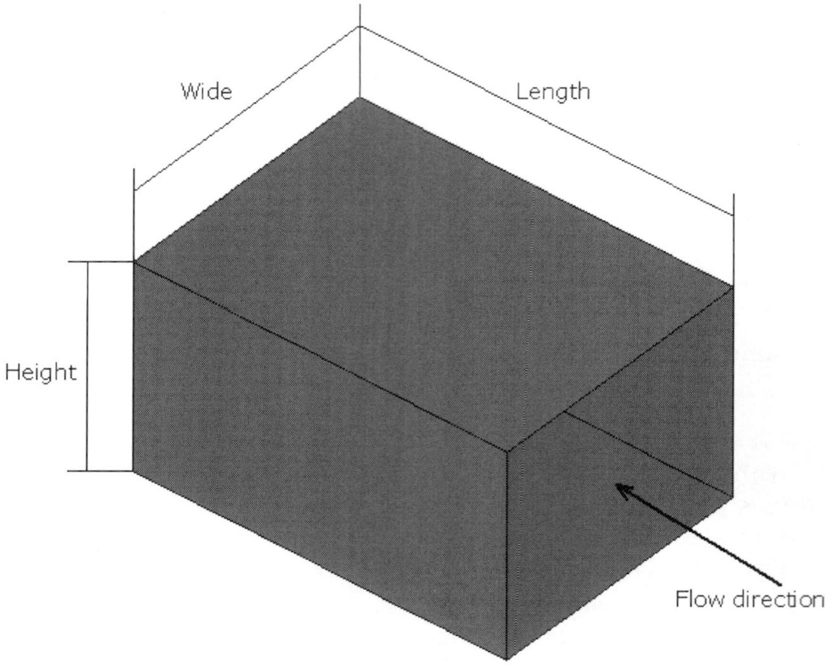

FIGURE 2. Layout of a constant section wind tunnel test chamber.

Figure 2 shows a design of a typical constant section test chamber. With the typical dimensions and velocities inside a wind tunnel, the flow in the test section, including the boundary layer, will be turbulent, because it is continuous along the whole wind tunnel. According to Idel´Cik (1969), the pressure loss coefficient, related to the dynamic pressure in the test section, which is considered as the reference dynamic pressure for all the calculations, is given by the expression:

$$\zeta=\lambda\cdot L/DH,$$

where L is the length of the test chamber, D_H the hydraulic diameter and λ a coefficient given by the expression:

$$\lambda=1/(1{,}8\cdot\log Re - 1{,}64)^2,$$

where Re is the Reynolds number based on the hydraulic diameter.

3.2. Contraction

The contraction or "nozzle" is the most critical part in the design of a wind tunnel; it has the highest impact on the test chamber flow quality. Its aim is to accelerate the flow from the settling chamber to the test chamber, further reducing flow turbulence and non-uniformities in the test chamber. The flow acceleration and non-uniformity attenuations mainly depend on the so-called contraction ratio, *N*, between the entrance and exit section areas. Figure 3 shows a typical wind tunnel contraction.

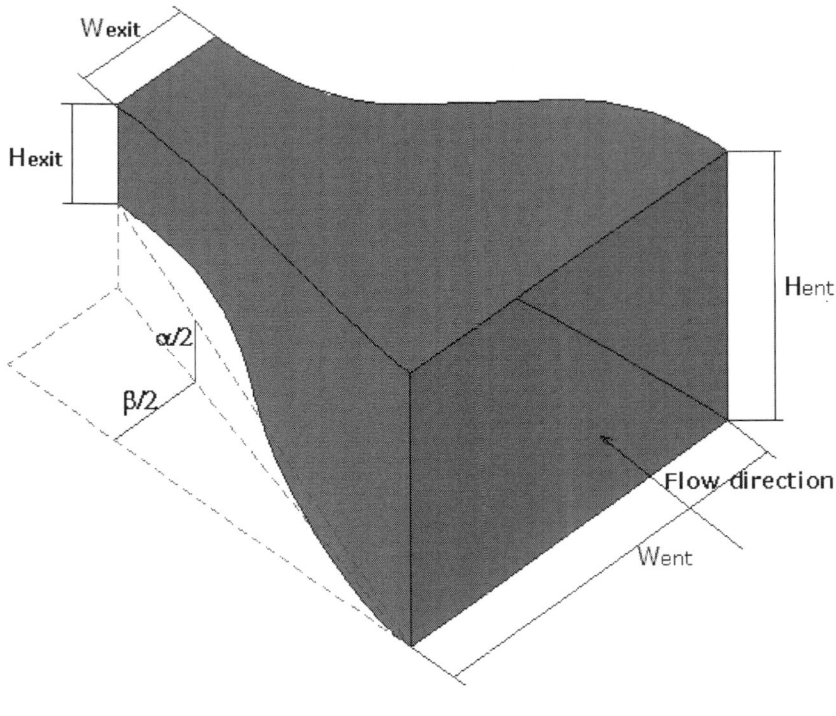

FIGURE 3. General layout of a three-dimensional wind tunnel contraction.

Although, due to the flow quality improvement, the contraction ratio, *N*, should be as large as possible, this parameter strongly influences the overall wind tunnel dimensions. Therefore, depending on the expected applications, a compromise for this parameter should be reached.

Quoting P. Bradshaw and R. Metha (1979), "The effect of a contraction on unsteady velocity variations and turbulence is more complicated: the reduction of x-component (axial) fluctuations is greater than that of transverse fluctuations. A simple analysis due to Prandtl predicts that the ratio of root-mean-square (rms) axial velocity fluctuation to mean velocity will be reduced by a factor $1/N^2$, as for mean-velocity variations, while the ratio of lateral rms fluctuations to mean velocity is reduced only by a factor of N: that is, the lateral fluctuations (in m/s, say) increase through the contraction, because of the stretching and spin-up of elementary longitudinal vortex lines. Batchelor, *The Theory of Homogeneous Turbulence*, Cambridge (1953), gives a more refined analysis, but Prandtl's results are good enough for tunnel design. The implication is that tunnel free-stream turbulence is far from isotropic. The axial-component fluctuation is easiest to measure, e.g. with a hot-wire anemometer, and is the "free-stream turbulence" value usually quoted. However, it is smaller than the others, even if it does contain a contribution from low-frequency unsteadiness of the tunnel flow as well as true turbulence."

In the case of wind tunnels for civil or industrial applications, a contractions ratio between 4,0 and 6,0 may be sufficient. With a good design of the shape, the flow turbulence and non-uniformities levels can reach the order of 2,0%, which is acceptable for many applications. Nevertheless, with one screen placed in the settling chamber those levels can be reduced up to 0,5%, which is a very reasonable value even for some aeronautical purposes.

For more demanding aeronautical, when the flow quality must be better than 0,1% in non-uniformities of the average speed and longitudinal turbulence level, and better than 0,3% in vertical and lateral turbulence level, a contraction ratio between 8,0 and 9,0 is more desirable. This ratio also allows installing 2 or 3 screens in the settling chamber to ensure the target flow quality without high pressure losses through them.

The shape of the contraction is the second characteristic to be defined. Taking into account that the contraction is rather smooth, one may think that a one-dimensional approach to the flow analysis would be adequate to determine the pressure gradient along it. Although this is right for the average values, the pressure distribution on the contraction walls has some regions with adverse pressure gradient, which may produce local boundary layer separation. When it happens, the turbulence level increases drastically, resulting in poor flow quality in the test chamber.

According to P. Bradshaw and R. Metha (1979), "The old-style contraction shape with a small radius of curvature at the wide end and a large radius at the narrow end to provide a gentle entry to the test section is not the optimum. There is a danger of boundary-layer separation at the wide end, or perturbation of the flow through the last screen. Good practice is to make the ratio of the radius of curvature to the flow width about the same at each end. However, a too large radius of curvature at the upstream end leads to slow acceleration and therefore increased rate of growth of boundary-layer thickness, so the boundary layer - if laminar as it should be in a small tunnel - may suffer from Taylor-Goertler "centrifugal" instability when the radius of curvature decreases".

According to our experience, when both of the contraction semi-angles, $\alpha/2$ and $\beta/2$ (see Figure 3), take the values in the order of 12º, the contraction has a reasonable length and a good fluid dynamic behaviour. With regard to the contour shape, following the recommendations of P. Bradshaw and R. Metha (1979), two segments of third degree polynomial curves are recommended.

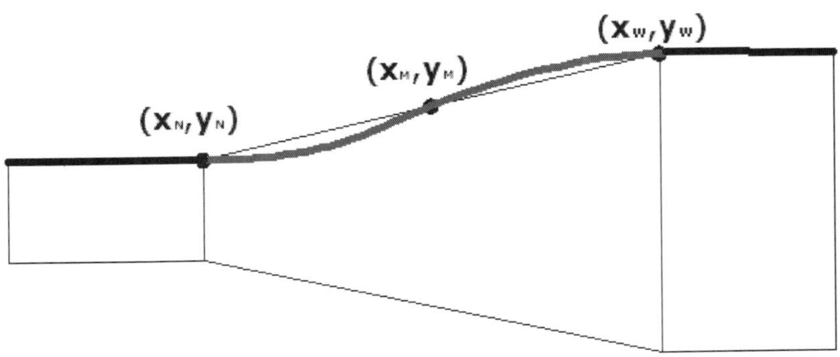

FIGURE 4. Fitting polynomials for contraction shape.

As indicated in Figure 4, the conditions required to define the polynomial starting at the wide end are: the coordinates (x_W, y_W), the horizontal tangential condition in that point, the point where the contour line crosses the connection strait line, usually in the 50% of such line, and the tangency with the line coming from the narrow end. For the line starting at the narrow end the initial point is (x_N, y_N), with the same horizontal tangential condition in this point, and the connection to the wide end line. Consequently, the polynomials are:

$$y = a_W + b_W \cdot x + c_W \cdot x^2 + d_W \cdot x^3,$$

$$y = a_N + b_N \cdot x + c_N \cdot x^2 + d_N \cdot x^3.$$

Imposing the condition that the connection point is in the 50%, the coordinates of that point are $[x_M, y_M] = [(x_W + x_N)/2, (y_W + y_N)/2)]$. Introducing the conditions in both polynomial equations, the two families of coefficients can be found.

According to Idel´Cik (1969), the pressure loss coefficient related to the dynamic pressure in the narrow section, is given by the expression:

$$\zeta = \left\{ \frac{\lambda}{\left[16 \cdot \sin\left(\frac{\alpha}{2}\right)\right]} \right\} \left(1 - \frac{1}{N^2}\right) + \left\{ \frac{\lambda}{\left[16 \cdot \sin\left(\frac{\beta}{2}\right)\right]} \right\} \left(1 - \frac{1}{N^2}\right),$$

where λ is defined as:

$$\lambda = 1/(1{,}8 \log Re - 1{,}64)^2.$$

The Reynolds number is based on the hydraulic diameter of the narrow section.

3.3. Settling Chamber

Once the flow exits the fourth corner (see Figure 1), the uniformization process starts in the settling chamber. In the case of low-quality flow requirements, it is a simple constant section duct, which connects the exit of the corner 4 with the entrance of the contraction.

Nevertheless, when a high quality flow is required, some devices can be installed to increase the flow uniformity and to reduce the turbulence level at the entrance of the contraction (see Figure 5). The most commonly used devices are screens and honeycombs. Both devices achieve this goal by producing a relatively high total pressure loss; however, keeping in mind that the local dynamic pressure equals to $1/N^2$ of the reference dynamic pressure, such pressure loss will only be a small part of the overall one, assuming that N is large enough.

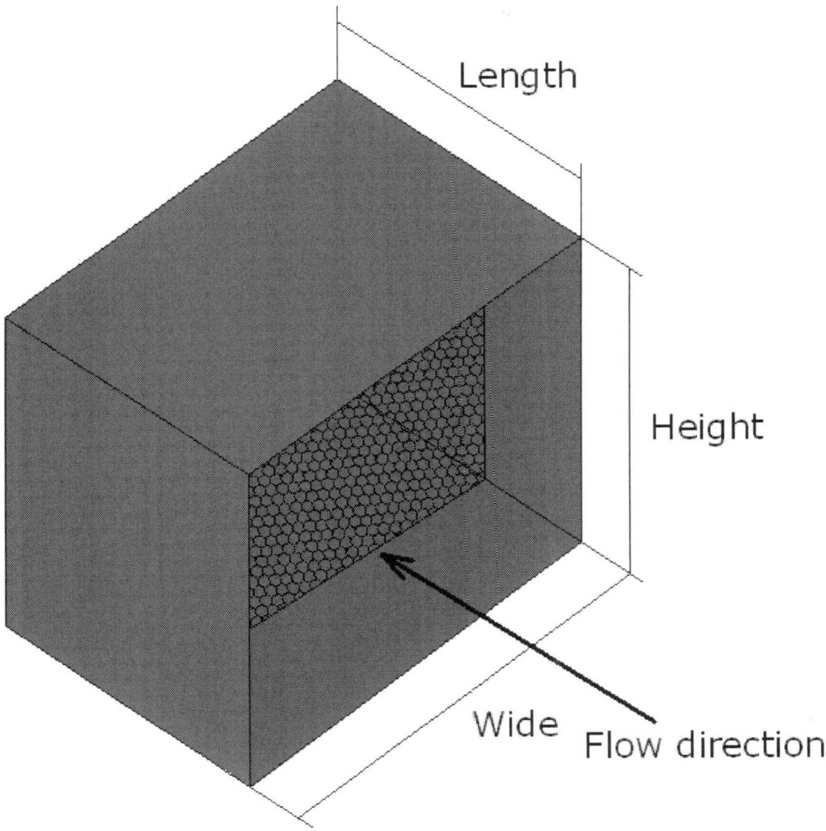

FIGURE 5. General layout of a settling chamber with a honeycomb layer.

Honeycomb is very efficient at reducing the lateral turbulence, as the flow pass through long and narrow pipes. Nevertheless, it introduces axial turbulence of the size equal to its diameter, which restrains the thickness of the honeycomb. The length must be at least 6 times bigger than the diameter. The pressure loss coefficient, with respect to the local dynamic pressure, is about 0,50 for a 3 mm diameter and

30 mm length honeycomb at typical settling chamber velocities and corresponding Reynolds numbers.

Although screens do not significantly influence the lateral turbulence, they are very efficient at reducing the longitudinal turbulence. In this case, the problem is that in the contraction chamber the lateral turbulence is less attenuated than the longitudinal one. As mentioned above, one screen can reduce very drastically the longitudinal turbulence level; however, using a series of 2 or 3 screens can attenuate turbulence level in two directions up to the value of 0,15%. The pressure loss coefficient, with respect to the local dynamic pressure, of an 80%-porous screen made of 0,5 mm diameter wires is about 0,40.

If a better flow quality is desired, a combination of honeycomb and screens is the most recommended solution. This configuration requires the honeycomb to be located upstream of 1 or 2 screens. In this case, the pressure loss coefficient, with respect to the local dynamic pressure, is going to be about 1,5. If the contraction ratio is 9, the impact on the total pressure loss coefficient would be about 0,02, which may represents a 10% of the total pressure loss coefficient. This implies a reduction of 5% in the maximum operating speed, for a given installed power.

The values of the pressure loss coefficients given in this section are only approximated and serve as a guideline for quick design decisions. More careful calculations are recommended for the final performance analysis following Idel´Cik's (1969) methods.

3.4. Diffusers

The main function of diffusers is to recover static pressure in order to increase the wind tunnel efficiency and, of course, to close the circuit. For that reason, and some other discussed later, it is important to maintain the flow attachment for pressure recovery efficiency. Figure 6 shows the layout of a rectangular section diffuser.

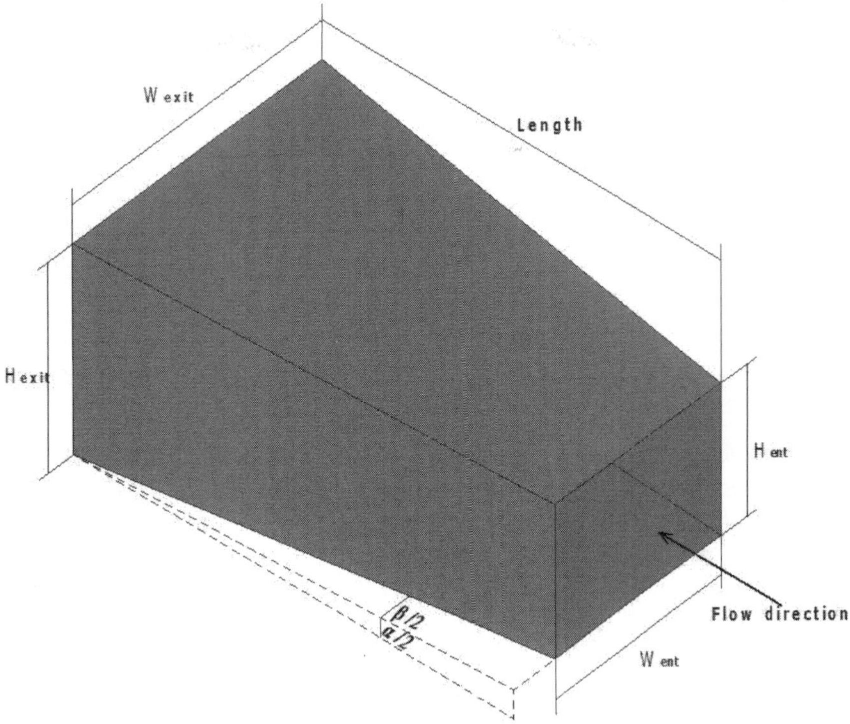

FIGURE 6. Rectangular section diffuser.

Diffuser 1 pays an important role in the test chamber flow quality. In case of flow detachment, the pressure pulsation is transmitted upstream into the test chamber, resulting in pressure and velocity non-uniformities. In addition, diffuser 1 acts as a buffer in the transmission of the pressure disturbances generated in the corner 1.

It has been proved that in order to avoid flow detachment, the maximum semi-opening angle in the diffuser has to be smaller than 3,5°. On the other hand, it is important to reduce as much as possible the dynamic pressure at the entrance of the corner 1, in order to minimise the possible pressure loss. Consequently, it is strongly recommended not to exceed the semi-opening angle limit and to design the diffuser to be as long as possible.

Diffuser 2 is a transitional duct, where the dynamic pressure is still rather large. Subsequently, the design criterion imposing a maximum value of the semi-opening angle must also be applied. The length of this diffuser cannot be chosen freely, because later it becomes restrained by the geometry of corners 3 and 4 and diffuser 5.

Diffuser 3 guides the flow to the power plant which is strongly affected by flow separation. In order to avoid it, the criterion imposing a maximum value of the semi-opening angle is maintained here as well. The cross-sectional shape may change along this diffuser because it must connect the exit of corner 2, whose shape usually resembles that of the test chamber, with the entrance of the power plant, whose shape will be discussed later.

The same can be said about diffuser 4 because pressure oscillations travel upstream and therefore may affect the power plant. Analogically to the previous case, it provides a connection between the exit of the power plant section and the corner 3, which has a cross-section shape resembling the one of the test chamber.

Diffuser 5 connects the corners 3 and 4. It is going to be very short, due to a low value of the dynamic pressure, which will allow reducing the overall wind tunnel size. This will happen mainly when the contraction ratio is high and the diffusion angle may be higher than 3,5°. It can also be used to start the adaptation between the cross-section shapes of the tests section and the power plant.

An accurate calculation of the pressure loss coefficient can be done with Idel´Cik´s (1969) method. A simplified procedure, derived from the method mentioned above, is presented here to facilitate a quick estimation of such coefficient.

The pressure loss coefficient, with respect to the dynamic pressure in the narrow side of the diffuser, is given by:

$$\zeta = 4{,}0 \cdot \tan \alpha/2 \cdot \sqrt[4]{\tan \alpha/2} \cdot \left(1 - \frac{F_0}{F_1}\right)^2 + \zeta_f .$$

α being the average opening angle, F_0 the area of the narrow section, F_1 the area of the wide section and where ζf is defined as:

$$\zeta_f = \frac{0{,}02}{8 \cdot \sin \alpha / 2}\left[1 - \left(\frac{F_0}{F_1}\right)^2\right].$$

3.5. Corners

Closed circuit wind tunnels require having four corners, which are responsible for more than 50% of the total pressure loss. The most critical contribution comes from the corner 1 because it introduces about 34% of the total pressure loss. To reduce the pressure loss and to improve the flow quality at the exit, corner vanes must be added. Figure 7 shows a typical wind tunnel corner, including the geometrical parameters and the positioning of corner vanes.

The width and the height at the entrance, W_{ent} and H_{ent} respectively, are given by the previous diffuser dimensions. The height at the exit, H_{exit}, should be the same as at the entrance, but the width at the exit, W_{exit}, can be increased, giving the corner an expansion ratio, W_{exit}/W_{ent}. This parameter can have positive effects on the pressure loss coefficient of values up to approximately 1,1. However, it must be designed considering specific geometrical considerations, which will be discussed, in greater details in the general arrangement.

The corner radius is another design parameter and it is normally proportional to the width at the corner entrance. The radius will be identical for the corner vanes. Although increasing the corner radius reduces the pressure loss due to the pressure distribution on corner vanes, it increases both the losses due to friction and the overall wind tunnel dimensions. According to previous experience, it is recommended to use 0,25 W_{ent} as the value of the radius for corners 1 and 2, and 0,20 W_{ent} for the other two corners.

The corner vanes spacing is another important design parameter. When the number of vanes increases, the loss due to pressure decreases, but the friction increases. Equal spacing is easier to define and sufficient for all corners apart from corner 1. In this case, in order to minimise pressure loss, the spacing should be gradually increased from the inner vanes to the outer ones.

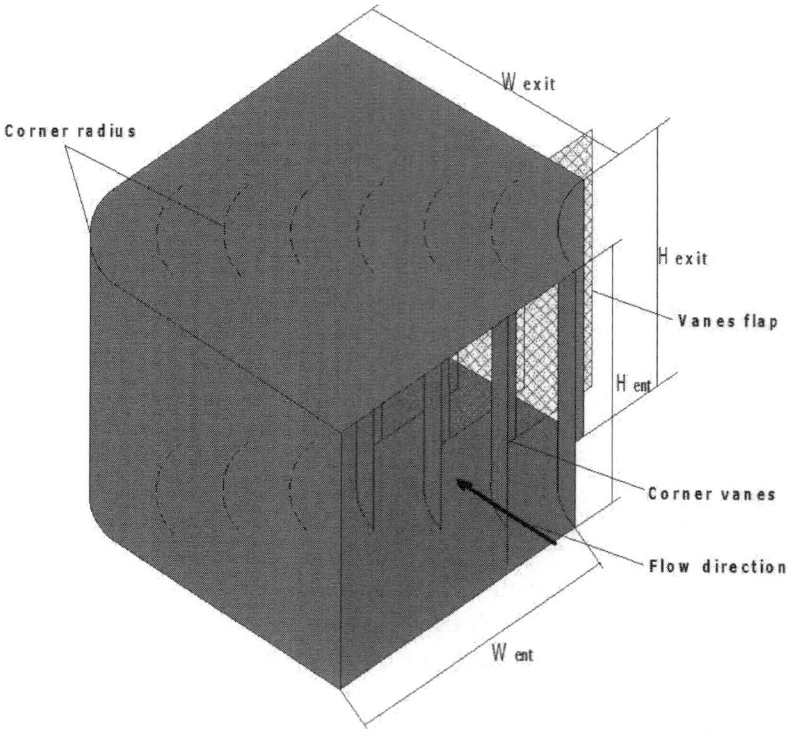

FIGURE 7. Scheme of a wind tunnel corner, including vanes, flaps and nomenclature.

The vanes can be defined as simple curved plates, but they can also be designed as cascade airfoils, which would lead to further pressure loss reduction. In the case of low speed wind tunnels the curved plates give reasonably good results. However, corner 1 may require to further stabilise the flow and reduce the pressure loss. Flap extensions with a length equal to the vane chord, as shown in Figure 7, is a strongly recommended solution to this problem.

Other parameters, such as the arc length of the vanes or their orientation, are beyond the scope of this chapter. For more thorough approach the reader should refer to Idel´Cik (1969), Chapter 6. As mentioned above, the pressure loss reduction in the corners is very important. Therefore, an optimum design of these elements, at least in the case of corner 1 and 2, has a significant impact on the wind tunnel performance.

In order to allow a preliminary estimation of the pressure loss in the corners we will follow the method presented in Diagram 6.33 from Idel´Cik (1969) mentioned above. In this approach, we take an average number of vanes, $n= 1{,}4{*}S/t_1$, S being the diagonal dimension of the corner, where t_1 is the chord of the vane. The pressure loss coefficient is given by the expression:

$$\zeta=\zeta_M+0{,}02+0{,}031{*}\frac{r}{W_{ent}}.$$

ζ_M depends on r/W_{ent}, and its values are 0,20 and 0,17 for r/W_{ent} equal to 0,20 and 0,25, respectively. As a result, the corresponding values of ζ are 0,226 and 0,198 respectively, always with respect to the dynamic pressure at the entrance. This proves the validity of the recommendations given before with regard to the value of the curvature radius and the length of diffusor 1.

3.6. Power Plant

The main aim of the power plant is to maintain the flow running inside the wind tunnel at a constant speed, compensating for all the losses and dissipation. The parameters that specify it are the pressure increment, Δp, the volumetric flow, Q, and the power, P. Once the test chamber cross-section surface, S_{TC}, and the desired operating speed, V, are fixed, and the total pressure loss coefficient, ζ, has been calculated, all those parameters can be calculated using:

$$\Delta p=\frac{1}{2}\rho\cdot V^2\cdot\zeta$$

$$Q=V\cdot S_{TC}$$

$$P=\Delta p\cdot\frac{Q}{\eta},$$

where ρ is the operating air density and η the fan efficiency, accounting for both aerodynamic and electric motor efficiencies.

In order to reduce the cost of this part by roughly one order of magnitude, we propose to use a multi-ventilator matrix, as presented in Figure 8, instead of a more standard single ventilator power plant configuration. The arrangement of this matrix will be discussed later.

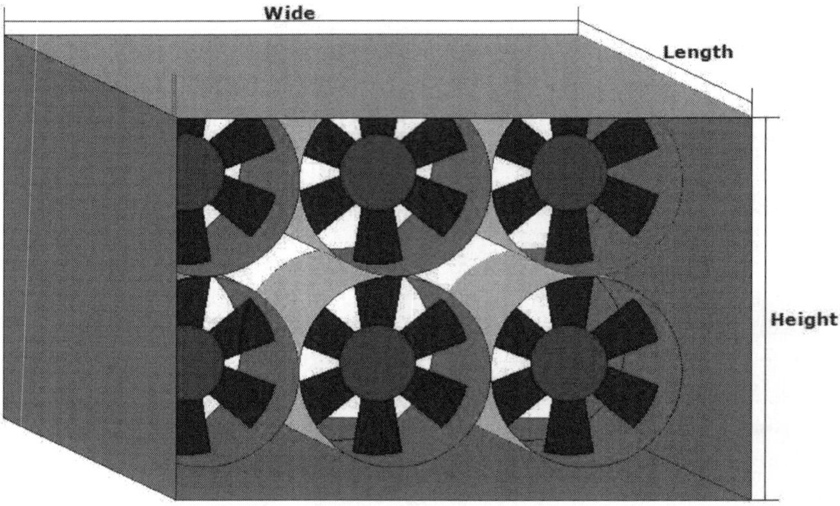

FIGURE 8. Layout of a multi-fan power plant.

According to our experience, for a closed circuit wind tunnel eventually including settling chamber screens or/and a honeycomb, the total pressure loss coefficient is in the range of 0,16 to 0,24. Consequently, in the case of 1,0 m^2 test section area and 80 m/s maximum operating speed, assuming an average value of ζ to be in the range mentioned above, and for a typical value of η equal to 0,65, the data specifying the power plant are:

$$\Delta p = 785 \text{ Pa}, Q = 80 \text{ m}^3/\text{s}, P = 100 \text{ kW}.$$

In this case we could use a 2,0m diameter fan specially designed for this purpose or 4 commercial fans of 1,0 m diameter, producing the same pressure increment, but with a volumetric flow of 20 m^3/s each. The latter option would reduce the total cost because the fans are a standard product.

4. GENERAL DESIGN PROCEDURE

The parameters that need to be defined in order to start the overall design are:

- Test chamber dimensions: width, W_{TC}, height, H_{TC}, and length, L_{TC}. These parameters allow to compute the cross-sectional area, $S_{TC}= W_{TC}\,H_{TC}$, and the hydraulic diameter, $D_{TC}=2\,W_{TC}H_{TC}/(W_{TC}+H_{TC})$.
- Contraction ratio, $N\approx5$ for low quality flow, and $N\approx9$ for high quality flow (considering the drawbacks of choosing a higher contraction ratio, explained before).
- Maximum operating speed, V_{TC}.

According to the impact on the wind tunnel dimensions and flow quality, Table 1 shows a classification of the design variables divided into two categories: main and secondary design parameters.

TABLE 1. Main and secondary wind tunnel design parameters

Main design parameters	Secondary design parameters
Maximum operating speed, V_{TC}	Contraction semi angle, $\alpha_C/2$
Test chamber width, W_{TC}	Settling chamber non-dimensional length, l_{SC}
Test chamber height, H_{TC}	Diffuser semi angle, $\alpha_D/2$
Test chamber length, L_{TC}	Diffuser 1 non-dimensional length, l_{D1}
Contraction ratio, N	Corner 1 expansion ratio, e_{C1}
	Corner 1 non-dimensional radius, r_{C1}
	Corner 4 non-dimensional radius, r_{C4}
	Diffuser 5 non-dimensional length, l_{D5}
	Corner 3 non-dimensional radius, r_{C3}
	Dimension of the fan matrix, n_W, n_H
	Unitary fan diameter, D_F
	Power plant non dimensional length, l_{PP}
	Corner 2 expansion ratio, e_{C2}

Now, following the guidelines given above, such as the convergence angle and the contour line shape of the contraction zone, the test and contraction chamber can be fully defined. In the case when both opening angles, α and β, are the same, the contraction length, L_C, is given by the expression:

$$L_C = \frac{(\sqrt{N} - 1) \cdot W_{TC}}{2 \cdot \tan(\alpha_C / 2)}.$$

Continuing in the upstream direction, the next part to be designed is the settling chamber. The only variable to be fixed is the length, because the section is identical to the wide section of the contraction. In the case when high quality flow is required, the minimum recommended non-dimensional length based on the hydraulic diameter, l_{SC}, is 0,60. This results from the necessity to provide extra space for the honeycomb and screens. In all other cases, the non-dimensional length may be 0,50. Therefore, the length of the settling, L_{SC}, chamber is given by:

$$L_{SC} = \sqrt{N} \cdot W_{TC} \cdot l_{SC}.$$

To obtain all the data for the geometric definition of the corner 4 satisfying all the recommendations given above we only need to fix the non-dimensional radius, r_{C4}. Its length, which is the same as its width, is:

$$L_{C4} = W_{C4} = \sqrt{N} \cdot W_{TC} \cdot (1 + r_{C4}).$$

Going downstream of the test chamber, we arrive at the diffuser 1. Assuming that both semi-opening angles are 3,5°, its non-dimensional length, l_{D1}, is the only design parameter. Although it has a direct effect on the wind tunnel overall length, we must be aware that this diffuser together with corner 1 are responsible for more than 50% of the total pressure losses. According to the experience, $l_{D1}>3$ and $l_{D1}>4$ is recommended for low and high contraction ratio wind tunnels respectively. The length of the diffuser 1, L_{D1}, and the width in the wide end, W_{WD1}, is defined by:

$$L_{D1} = W_{TC} \cdot l_{D1}$$

$$W_{WD1} = [1 + 2 \cdot l_{D1} \cdot \tan(\alpha_{D1}/2)] \cdot W_{TC}.$$

With regard to the corner 1, once its section at the entrance is fixed (it is constrained by the exit of diffuser 1), we must define the non-dimensional radius, r_{C1}, and the expansion ratio, e_{C1}. As a result, the width at the exit, W_{EC1}, the overall length, L_{C1}, and width, W_{C1}, can be calculated using:

$$W_{EC1} = W_{WD1} \cdot e_{C1}$$

$$L_{C1} = W_{WD1} \cdot (e_{C1} + r_{C1})$$

$$W_{C1} = W_{WD1} \cdot (1 + r_{C1}).$$

Therefore, we can already formulate the overall wind tunnel length, L_{WT}, as a function of the test chamber dimensions, the contraction ratio, and other secondary design parameters:

$$L_{WT} = L_{TC} + W_{TC} \cdot \left[\frac{(\sqrt{N} - 1)}{2 \cdot \tan(\alpha_C/2)} + \sqrt{N} \cdot l_{SC} + \sqrt{N} \cdot (1 + r_{C4}) + l_{D1} + [1 + 2 \cdot l_{D1} \cdot \tan(\alpha_{D1}/2)] \cdot (e_{C1} + r_{C1})\right].$$

This quick calculation allows the designer to check whether the available length is sufficient to fit the wind tunnel.

Taking into account all the recommended values for the secondary design parameters, a guess value for the wind tunnel overall length, with a contraction ratio N=9 (high quality flow), is given by the formula:

$$L_{WT} = L_{TC} + 16 \cdot W_{TC}.$$

In the case when N=5 (low quality flow), the formula becomes:

$$L_{WT} = L_{TC} + 11{,}5 \cdot W_{TC}.$$

The designer must be aware that any modification introduced to the secondary design parameters modifies only slightly the factor that multiplies W_{TC} in the formulas above. Consequently, if the available space is insufficient, the only solution would be to modify the test chamber dimensions and/or the contraction ratio.

As we have already defined the wind tunnel length using the criterion of adequate flow quality, we can now devote our attention to designing the rest of the circuit, the so-called return circuit. The goal is not to increase its length, intending also to minimize the overall width and keeping the pressure loss as low as possible.

Keeping this in mind, the next step in the design is to make a first guess about the power plant dimensions. Following our design recommendations, a typical value for the total pressure loss coefficient of a low contraction ratio wind tunnel, excluding screens and honeycombs in the settling chamber, is 0,20, with respect to the dynamic pressure in the test chamber. This value is approximately 0,16 for a large contraction ratio wind tunnels. If screens and honeycombs were necessary, those figures could increase by about 20%.

As the power plant is placed more or less in the middle of the return duct, the area of the section will be similar to the mid-section of the contraction. Therefore, taking into account the volumetric flow, the total pressure loss, and the available fans, the decision about the type of fan and the number of them can be taken. Using this approach, the power plant would be defined, at least in the preliminary stage.

We will return now to the example we started before for the power plant section. To improve the understanding of the subject, we are going to present a case study. If the test chamber section was square and N=5, the mid-section of the contraction would be 1,67 x 1,67 m^2. This would allow us to place 4 standard fans of 0,800 m diameter each. The maximum reduction in the width size would be obtained by suppressing the diffuser 5, obtaining the wind tunnel platform shown in Figure 9. We have not defined the diffusion semi-angle in diffuser 3, but we checked afterwards that it was smaller than 3,5°. Figure 9 is just a wire scheme of the wind tunnel, made with an Excel spreadsheet, and for this reason the corners have not been rounded and are represented just as boxes.

In the case of a 4:3 ratio rectangular test chamber cross-section, the mid-section of the contraction would be 1,869x1,401 m^2 and for this reason we could suggest the use of 6 standard fans of 0,630 m diameter, organized in a 3x2 matrix, occupying a section of 1,890x1,260 m^2. Figure 10 shows the wire scheme of this new design. We can check that the diffuser 3 semi-angle is below 3,5° as well.

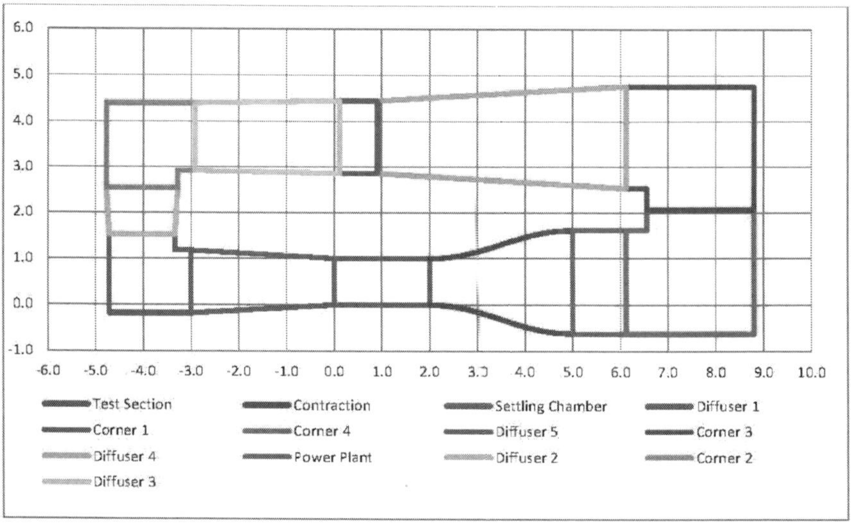

FIGURE 9. Non-dimensional scheme of a wind tunnel with square section test chamber and low contraction ratio, N≈5.

It is clear that the new design is slightly longer and wider, but it is because of the influence of the test chamber´s width, as shown above.

Notice that in both cases corner 3 has the same shape as corner 4. Similarly, the entrance section of diffuser 4 is the same as of the power plant section, and using a diffuser semi-angle of 3,5°, this item is also well defined.

At this stage we have completely defined the wind tunnel centre line, so that we can calculate the length, L_{CL}, and width, W_{CL}, using:

$$L_{CL} = (L_{C1} - W_{EC1}/2) + L_{D1} + L_{TC} + L_{C} + L_{ST} + (L_{C4} - W_{ED5}/2)$$

$$W_{CL} = (W_{C4} - W_{EC4}/2) + L_{D5} + (W_{C3} - W_{ED4}/2).$$

The distance between the exit of the corner 1 and the centre of the corner 2, DC1_CC2, can be calculated through the expression (see Figure 11):

$$D_{C1\_CC2} = W_{CL} - W_{ED1}\left(r_{C1} + \frac{1}{2}\right).$$

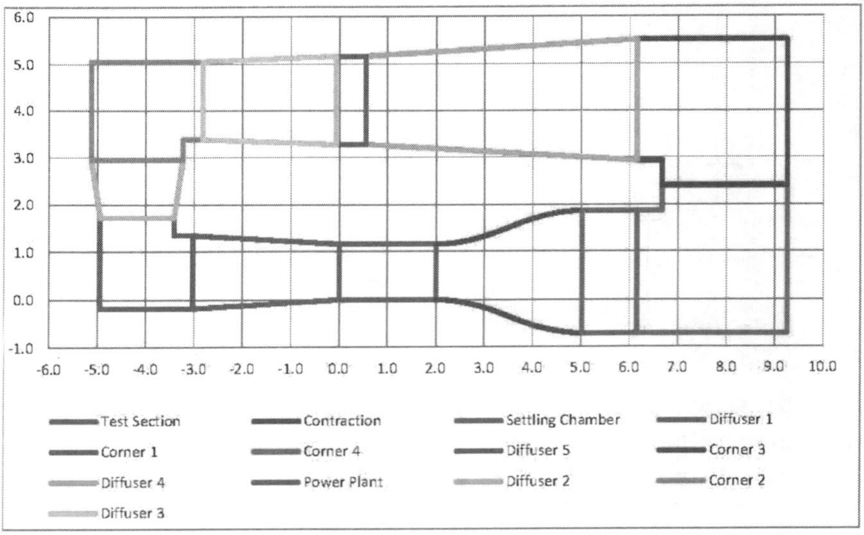

FIGURE 10. Non-dimensional scheme of a wind tunnel with rectangular section test chamber and low contraction ratio, N≈5.

On the other hand:

$$D_{C1\_CC2} = L_{D2} + [(W_{C2} - W_{EC2}/2)]$$

$$W_{EC2} = W_{ED2} \cdot e_{C2}$$

$$W_{C2} = W_{ED2} \cdot (r_{C2} + e_{C2})$$

$$W_{ED2}=W_{EC1}+2\cdot L_{D2}\cdot \tan(\alpha_{D2}/2).$$

Manipulating and combining those equations, we obtain:

$$L_{D2}=\frac{D_{C1\_CC2}-W_{EC1}\cdot(r_{C2}+e_{C2}/2)}{1+2\cdot(r_{C2}+e_{C2}/2)\cdot\tan(\alpha_{D2}/2)}.$$

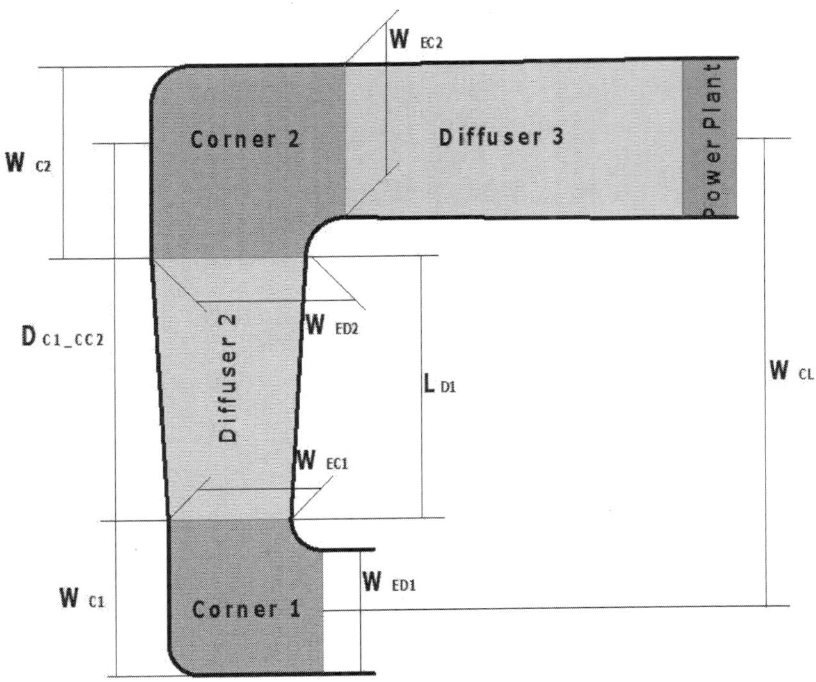

FIGURE 11. Scheme with the definition of the variable involving the design of diffuser 2 and 3, and corner 2.

With this value, by substituting it into the previous expressions, we have all the parameters to design diffusers 2 and 3, and corner 2. Finally, it is necessary to check that the opening angles of diffuser 3 are below the limit. In case when the vertical opening angle, α, exceeds the limit, the best option is to increase the diffuser 1 length, if this is possible, because it improves flow quality and reduces pressure loss. If the wind tunnel length is in the limit, another option is to add the diffuser 5 to the original scheme. However, it will increase the overall width. When the limit of the horizontal opening angle, β, is exceeded, then the best option is to adjust the values of the expansion ratio in corners 1 and 2, because it will not change the overall dimensions.

The following case study is a wind tunnel with high contraction ratio, $N \approx 9$, and square section test chamber. In this case, the approximate area of the power plant section will be 2,000 x 2,000 m^2. In this case we have two compatible options to select the power plant. We can just select a matrix of 4 fans, 1,000 m diameter each. However, if the operating speed is rather high, in order to be able achieve the required pressure increment and the mass flow, we may need to use 1,250 m diameter fans. Figure 12shows both options. Note that the overall planform is only slightly modified and the only difference is the position where the power plant is placed.

The design of the diffusers 2 and 3, and the corner 2 will be done following the same method as for the previous cases.

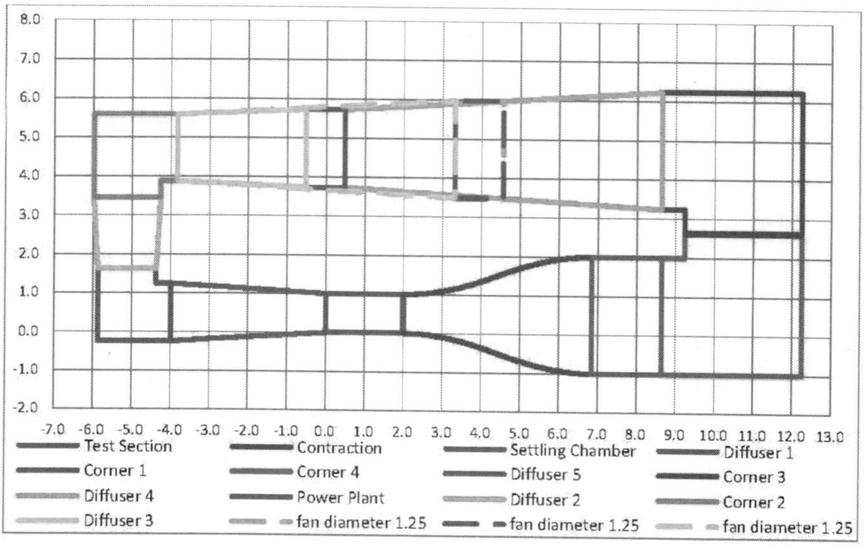

FIGURE 12. Non-dimensional scheme of a wind tunnel with square section test chamber and high contraction ratio, N≈9. Two different standard power plant options are presented.

5. WIND TUNNEL CONSTRUCTION

One of the most important points mentioned in this chapter refers to the wind tunnel cost, intending to offer low cost design solutions. Up to now we have mentioned such modifications to the power plant, proposing a multi-fan solution instead of the traditional special purpose single fan.

The second and most important point is the wind tunnel's construction. The most common wind tunnels, including those with square or rectangular test sections, have rounded return circuits, like in the case of the NLR-LSWT. However, the return circuit of DNW wind tunnel is constructed with octagonal sections. Although the second solution is cheaper, in both cases different parts of the circuit needed to be built in factories far away from the wind tunnel location, resulting in very complicated transportation operation.

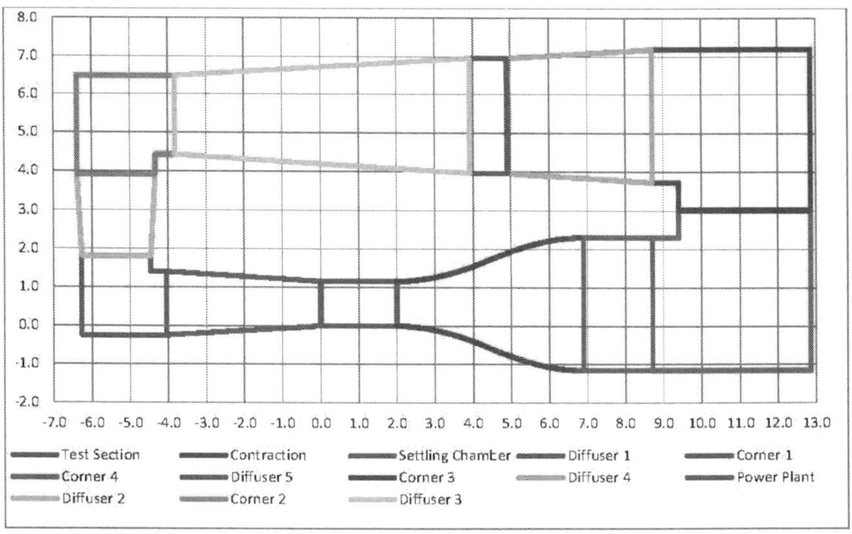

FIGURE 13. Non-dimensional scheme of a wind tunnel with rectangular section test chamber and large contraction ratio, N≈9.

To reduce the costs, all the walls can be constructed with flat panels, which can be made on site from wood, metal or even concrete, like in the case of ITER's wind tunnel. Figure 14 shows two wind tunnels built with wood panels and aluminium standard profile structure.

Both wind tunnels shown in Figure 14 are open circuit. The one on the left is located in the Technological Centre of the UPM in Getafe (Madrid) and its test chamber section is 1,20 x 1,00 m^2. Its main application is mainly research. The right one is located in the Airplane Laboratory of the Aeronautic School of the UPM. Its test chamber section is 0,80 x 1,20 m^2, and it is normally used for teaching purposes, although some research projects and students competitions were done there as well. Despite the fact that these tunnels are open circuit, the construction solutions can be also applied to closed circuit ones.

FIGURE 14. Research and education purpose wind tunnels built with wood panel and standard metallic profiles, with multi-fan power plant.

According to our experience, the manpower cost to build a wind tunnel like those defined in figures 9to 13 could be 3 man-months for the design and 16 man-months for the construction. With these data, the cost of the complete circuit, excluding power plant, would be about 70.000,00 €. In our opinion, the cost figure is very good, considering the fact that the complete building time possibly may not exceed even 9 months.

We have more reliable data with regard to the ITER wind tunnel, built in 2000-01. The whole cost of the wind tunnel, including power plant, work shop and control room, was 150.000,00 €.

This wind tunnel was almost completely built with concrete. Figure 15 shows different stages of the construction, starting from laying the foundations to the almost final view. The small photos show the contraction, with the template used for wall finishing, and the power plant.

FIGURE 15. Photographic sequence of the construction of the ITER Low Speed Wind Tunnel. The top left picture shows the foundations, the top right the contraction, the bottom left the power plant and the bottom right a view from the outside almost at the end of the construction.

6. CONCLUSIONS

A method for quick design of low speed and low cost wind tunnels, either for aeronautical and/or civil applications, has been presented.

The possibility of deciding between both applications means that the method allows achieving flow quality level as good as desired.

The method also allows to the designer to get a quick and rough estimation of the overall wind tunnel size, once the main design parameters are given.

The guidelines to choose the secondary design parameters are given as well.

To address the low cost of design and construction, the use of a multi fan power plant and rectangular duct sections is proposed as well.

NAN. ACKNOWLEDGEMENTS

The authors would like to acknowledge to Instituto Tecnológico y de Energías Renovables (ITER) and to Grupo λ_3 of the UPM for their contribution.

REFERENCES

1. J. B Barlow, W. H Rae, A Pope, Low-speed wind tunnel testing, John Wiley & Sons New York, 1999rd ed.
2. G. G Borger, The optimization of wind tunnel contractions for the subsonic range, NASA Technical Translation / F-16899, NASA Washington, 1976
3. W. T Eckert, K. W Mort, J Jope, Aerodynamic design guidelines and computer program for estimation of subsonic wind tunnel performance, NASA technical note / D-8243, NASA Washington, 1976
4. S. M Gorlin, I. I Slezinger, Wind tunnels and their instrumentation, Israel Program for Scientific Translations Jerusalem, 1966
5. Idel'Cik. I.E., Memento des pertes de charge: Coefficients de pertes de charge singulières et de pertes de charge par frottement, Eyrolles Editeur, Paris 1969

6. E. C Maskell, A theory of the blockage effects on bluff bodies and stalled wings in a closed wind tunnel, R. & M. 3400, November, 1963

7. R. D Mehta, and P Bradshaw, Design Rules for Small Low-Speed Wind Tunnels, Aero. Journal (Royal Aeronautical Society), 734431979

8. J Scheiman, Considerations for the installation of honeycomb and screens to reduce wind-tunnel turbulence, NASA Technical Memorandum / 81868, NASA Washington, 1981

9. The Royal Aeronautical Society. Wind tunnels and wind tunnel test techniques, Royal Aeronautical Society London, 1992

2

CHAPTER

Design Optimization of Three Dimensional Geometry of Wind Tunnel Contraction

A.S. Abdelhamed , Y.El-S. Yassen, , M.M. ElSakka

ABSTRACT

The aim of the present study is to redesign three dimensional geometry of existing open circuit wind tunnel contraction. The present work achieves the recommended contraction ratio, maximum uniformity at the working section mid-plane, without separation, no Gortler vortices in the contraction, and minimizing the boundary layer thickness at entrance to the working section. Using CFD along with optimization tools can shorten the design optimization cycle time. Moreover CFD allows insight into the minute flow details which otherwise are not captured using flow bench tests. The design exploration algorithm is used to optimize the profile of the contraction in an automated manner. The optimization is based on using screening method to choose the best design set and verified by the CFD solver. The new contraction, compared to the old design contraction is confirmed using CFD. The new design is manufactured in full scale. The optimized contraction is investigated computationally and experimentally.

Keywords: Wind tunnel contraction; Turbulence model; Goal-driven optimization; Screening method; Uniformity; Analytical technique

1. INTRODUCTION

Contraction is an important part of a wind tunnel. The main effects of a contraction are to reduce both mean and fluctuating velocity variations to a smaller fraction of the average velocity and further to increase the corresponding mean velocity. Generally to design of a subsonic and supersonic wind tunnel, the contraction portion should not have adverse pressure gradient in the stream-wise and further the effect of adverse pressure gradient at the exit of the contraction must be minimal. Whenever a converging duct segment is attached to constant-area segments, regions of adverse pressure gradient will occur along the wall, at its inlet and exit that may cause boundary layer separation. If separation occurs, it will degrade the flow uniformity and steadiness, both of which are essential in a test facility. Separation is usually avoided if the adverse pressure gradients are minimized which is done by making the contraction sufficiently long. The contraction can be divided into two parts. The first part has walls of concave shape and it is very important to elongate this part as much as possible to avoid wall boundary layer separation. The streamline curvature effects on the pressure gradient in the boundary layer promote the risk of separation. Along a fair part of this section, there will be a positive pressure gradient. The second part of the contraction has convex walls that may cause flow separation in the vicinity of the test section due to a positive pressure gradient. To delay separation, it is better that a longer contraction's length be chosen, but this will increase the cost and thickening the boundary layer that may enhance boundary layer and risk of separation. Furthermore, if the length is reduced, the contraction costs will reduce and it will fit into a smaller space. In addition, the boundary layer will generally be thinner due to the combined effects of increase in the favorable pressure gradients and decrease in the length of the contraction. Furthermore, it may increase the possibility of flow separation. Thus the length most be optimized. The contraction area ratio is another dominant factor that affects the extent of flow uniformity, flow separation, and downstream turbulence level. The characteristics of the flow were investigated in a square contraction numerically and experimentally [1]. Their Measurements included the cross-sectional velocity profiles and longitudinal pressure distributions along the wall of a contraction of a wind tunnel. Boundary layer was studied in the contraction [2]. The calculations showed that the minimum-length contraction shapes can provide fully attached boundary-layer flow. The results showed that exist regions of separated flow along the wall, in the inlet and exit of the contraction furthermore; conclusions were drawn that the existing adverse pressure gradient is the essential condition for the separation. Flow study shows that a three-dimensional separation occurs in the contraction surface. In a proposed conceptual model of this phenomenon, the separation process begins with small non-uniformities in the boundary-layer flow merging from the screens upstream of the contraction. On entering the contraction, the non-uniformities are amplified by a combination of Gortler instability, lateral pressure gradient and adverse streamwise pressure gradient to form a strong counter-rotating streamwise vortex pair (Gortler vortex) that detaches from the surface [3] and [4].

Another contraction parameter that has to be selected, a priori, is the cross-sectional shape. In order to avoid cross-flows and boundary layer separation in the corners, the ideal cross-sectional shape is circular. However, in the absence of separation, the secondary flows in the corners tend to remain localized, without any significant effect on the test section flow quality [5]. The cross-sectional shape for modern day contractions is, therefore, almost always chosen to match the other tunnel components which are normally square or rectangular.

An iterative design procedure was developed for the contraction to be installed on the mixing layer wind tunnel [6] and [7]. The procedure consisted of first computing the potential flow field and hence the pressure distributions along the walls of a contraction of given size and shape using a three-dimensional numerical panel method.

Currently, more flexibility in the design of wind tunnel contractions can be exhibited, with the use of CFD to enable rapid testing of designs to optimize contractions of arbitrary cross-section and wall profile. The use of CFD allows for the use of design exploration algorithm to optimize the profile of the contraction in an automated manner. However, the performance of the contraction still requires testing after construction, as the level of CFD used for this application is typically insufficient to detect the development of longitudinal vortices through the working section such as were measured by [8].

A great effort has been made to set the guidelines for the design of wind tunnels[9] and [10]. However, the recent development in optimization techniques encourages the researchers to develop optimized design for the wind tunnels especially the wind tunnel contraction that has a strong influence on the flow quality in the test section. An optimization is performed for a 2D contraction profile described by a six order polynomial and the results are validated experimentally [11]. The effective Global Optimization algorithm is used to optimize the profile of a 2D contraction described by a two parameters Bézier curve modeled by a three- dimensional potential flow solver using Kriging Meta model to predict the values of the objective function [12].

The present work contains a detailed account of the redesign of the existing contraction that based on the analytical technique according to the method given by [13] and [14], assembly and calibration of a wind tunnel specifically redesigned for maximum flow uniformity at the working section, prevention of separation and no Gortler vortices in the contraction, controlling the turbulence level and minimizing the boundary layer thickness at entrance to the working section. The wind tunnel consists of two separate legs which are driven independently by centrifugal blowers connected to variable speed motors. The two streams are allowed to merge in the test section. The construction of the wind tunnel was motivated by a strong interest in the study of three dimensional contraction profiles. The wall contraction profile is described by three dimensional delineated curves and optimized using the design exploration algorithm based on screening method. The flow in the wind tunnel is modeled using a Reynolds-averaged Navier–Stokes (RANS) solver. The aerodynamic performance improvement of the optimized contraction, compared to the old design contraction based on analytical technique, is confirmed using CFD. The new design is manufactured in full scale. Although, investigation by using CFD analysis is less expensive in general, the results obtained from CFD calculations should be validated by means of experimental results. In addition to validation, in cases like simulating three dimensional contraction profiles, performing CFD simulations can be time consuming. Therefore, it is essential to find approaches that can reduce the computational cost while results are validated by experiments. The performance of the optimized contraction is investigated computationally and experimentally.

2. WIND TUNNEL DESCRIPTION

The wind tunnel facility included in this work is an open jet blower type driven by two similar 5.5 kW centrifugal blowers. Each blower supplies air into a separated branch through a flexible joint. Each branch consists of two wide angle diffuser connected in series. The first diffuser section expands from 36 × 50 cm to 50 × 100 cm over a length of 125 cm, giving a maximum angle of 22.6° and an area ratio of 2.78. The second diffuser expands from 50 × 100 cm to 74 × 150 cm over a length of 100 cm, giving a maximum angle of 28° and an area ratio of 2.22. A screen with a porosity of 0.43 is placed between the two diffusers while a screen with a porosity of 0.5 is placed at the exit of the second diffuser. The flow from the two branches is merged in the settling chamber. The merged flow passes through the contraction section as shown in Fig. 1.

The old contraction has a square cross section of 150 × 150 cm^2 at inlet, 80 × 80 cm^2 at exit, and a contraction ratio (CR) of 3.5 that is lower than the recommended contraction ratio (6:9) [13] and [14]. The optimized contraction has an exit cross section of 60 × 60 cm^2 and a contraction ratio of 6.25 over a length (L) of 150 cm that gives the recommended ratio of unity between length and hydraulic diameter at the contraction inlet [14]. The two geometries of old and optimized contractions are shown in Fig. 2.

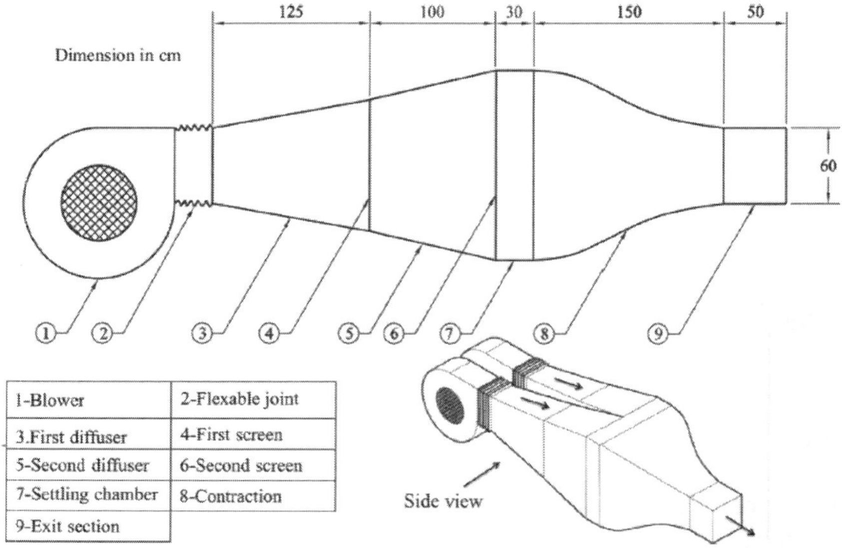

Figure 1. Wind tunnel.

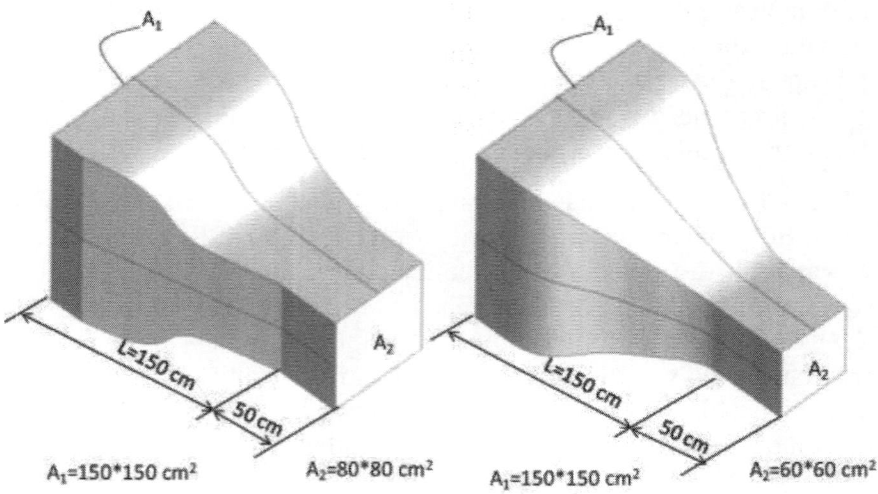

Figure 2. Shape of the old contraction (left) and the new design (right).

3. EXPERIMENTAL PROCEDURE

The geometry of optimized contraction was manufactured in full scale. In order to achieve the inlet boundary condition (12 m/s) compared to the CFD calculation, the two similar 5.5 kW centrifugal blowers are to be controlled by an electrical inverter. The velocity profile at exit section of the contraction is measured using Static-Pitot tube of 6 mm diameter. A simple slide probe traversing mechanism is used to locate the probe tip at every location (30 points, and 1 cm between each two successive points and three points, 2 mm between each near the wall except the nearest to which is at 4 mm from the wall). The collected measurements data at measuring station locations were averaged and used to calculate the location mean velocity distribution of the air flow through the duct, for Reynolds number ($Re = 4.8 \times 10^5$). The uncertainty, using root-sum-square method in the measured mean velocity is ±4.8% within 95% confidence.

4. DESIGN EXPLORATION AND OPTIMIZATION

The optimization process is performed using the Design Exploration algorithm described in Fig. 3. The optimization is the process to choose the optimal input parameters that achieve the desired performance of the system in an automated manner. This performance is evaluated by means of output parameters. There are several choices to examine the system performance using a certain output parameter such as maximize its value, minimize its value, or seeking target value.

Initially, the complete analysis of an initial design has to be held including drawing, meshing, solving the governing equations, defining the input and output parameters.

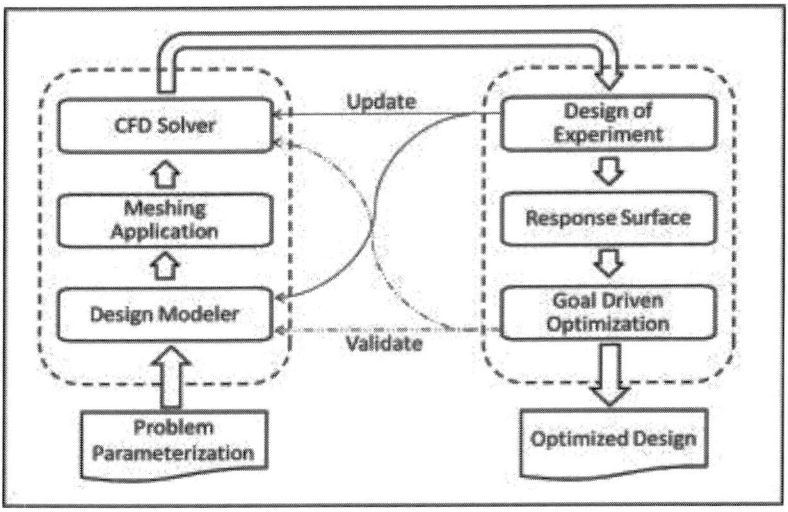

Figure 3. Design exploration algorithm.

4.1. Problem parameterization

The profile of the wind tunnel contraction consists of concave and convex parts. The method of the delineation of the profile is based on the perception of its variation from a straight line [15]. This method has been utilized to parameterize the profile of the contraction into three dimensionless parameters *a*, *b*, and *c*. These parameters are defined in the Design Modeler to describe the profile and the variation of the profile of the contraction as shown in Figure 4 and Figure 5. The objective function of the optimization process is to maximize the flow uniformity at the exit plane of the contraction, $Uni_{ex}$where:

$$Uni_{ex} = \left| \frac{U_{max}}{U_{mean}} \right|_{exit\ plane}$$

The minimum positive wall shear on the contraction is used to attribute the absence of separation in the contraction.

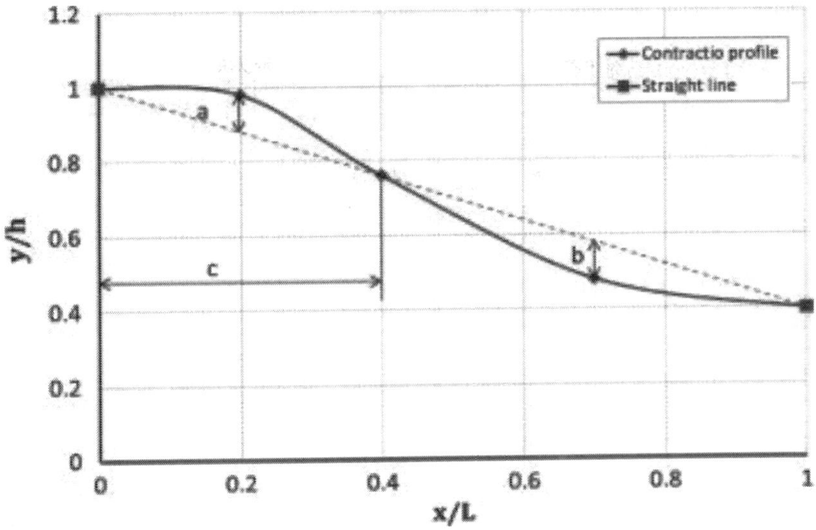

Figure 4. The contraction profile parameterization.

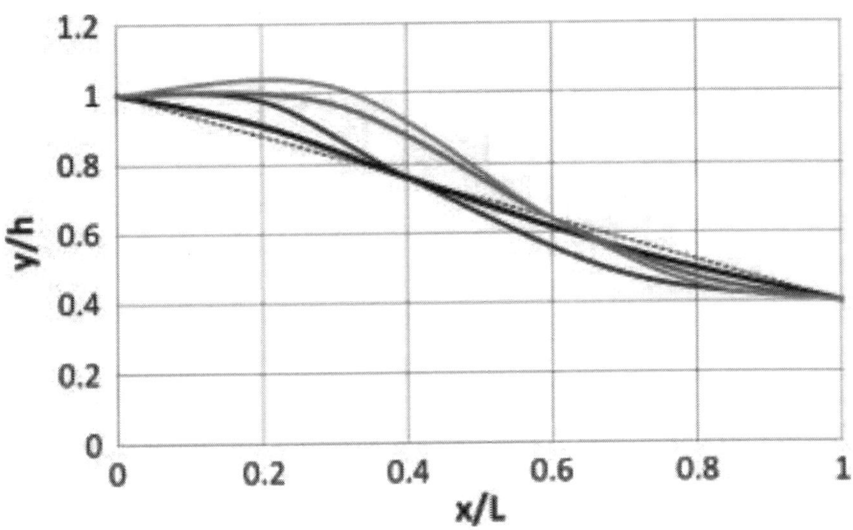

Figure 5. The variation of the profile with dimensionless parameters *a*, *b*, *c*.

4.2. Computational treatment

The CFD solver used for this study is a Fluent (ANSYS 14) Pressure Based Segregated Solver. Quadratic Upwind Interpolation (QUICK) that is a 3rd-order accurate is used for spatial discretization. Shear Stress Transport (SST – k–ω) model is used for turbulence modeling. One quarter of the wind tunnel is modeled, using the horizontal and vertical symmetry planes to reduce computational time. The domain is meshed using a hexahedral mesh. The mesh of the final design shown in Fig. 6 has 254,259 elements with a maximum aspect ratio of 2.7778 and average skewness factor of 0.089395. The extensive numerical simulation illustrates that the hexahedral mesh is more stable and robust to converge. In spite of good results, usage of this type makes the mesh not fine enough near the wall to resolve the characteristics of the boundary layer. The inlet boundary condition to the domain is 12 m/s. The outlet boundary condition is assumed to be constant static atmospheric pressure at exit.

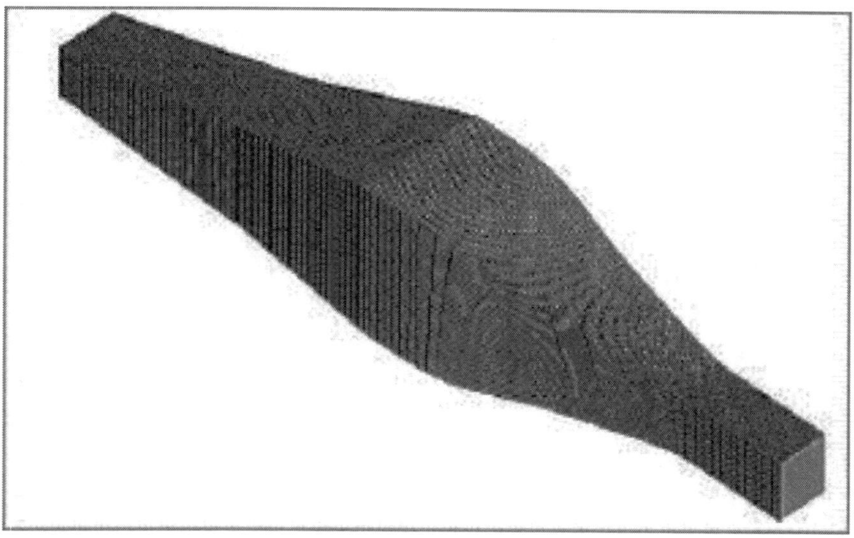

Figure 6. Hexahedral mesh of the final design one quarter of the wind tunnel (grid representation).

The screens are represented using the porous jump model with a pressure jump coefficient of C_2 = 218.875 [1/m] for the first screen and 142 [1/m] for the second one. These coefficients are derived by the relation provided by [16] and [17] for the pressure drop coefficient $k = 0.71 * (1 - \beta)/\beta^2$.

4.3. Goal Driven Optimization

The Goal Driven Optimization is performed in three steps:

4.3.1. Design of Experiment (DOE)

In this step a test sample of the input parameters is generated to cover the whole range of these parameters with minimum number of design points, keeping this sample efficiently represents the effects of the input parameters on the output parameters. DOE technique tries to locate the sampling points in the design space such that the space for random input parameters is explored in most efficient way, or to obtain the required information with a minimum of sampling points [18].

After generation of the test sample in the DOE table using one of DOE methods. The update process is performed in which the optimizer repeats all the steps have been done in the initial design (drawing/meshing/solving generating output parameters) for all the design points in the DOE table.

4.3.2. Response surface

In this step the results generated in DOE table is utilized to build a multidimensional correlation. This correlation will be used to predict the response or the behavior of the system to any change of the input parameters. The predicted response surface is used to get the value of the output variables to any combination of the input parameters without the need to perform a complete simulation.

4.3.3. Optimization

In this step the predicted response surface is utilized to choose the optimal input parameters that achieve the desired performance or characteristics of the tested system using a comparative study by checking the output parameters to be maximized, minimized, seeking target value

In this study, the initial design has been setup. The Design of Experiments (DOE) technique is used for sampling. The Enhanced-Rotatable Central Composite Design (CCD) is chosen for DOE that is a five level sampling design as the other three level sampling methods is insufficient to describe the problem. Fig. 7 shows the parameters parallel chart for 29 sample points with input parameters.

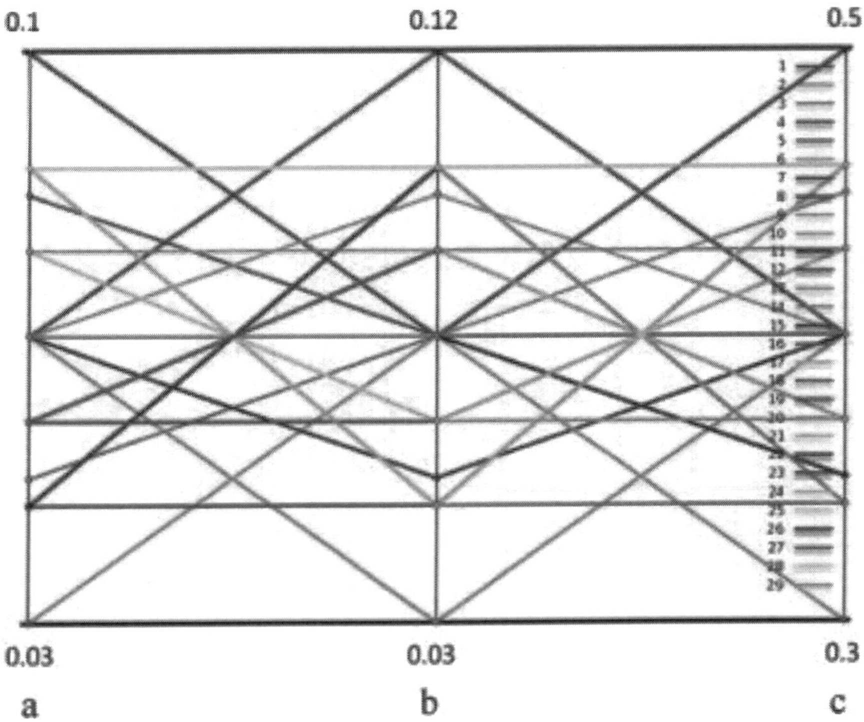

Figure 7. Parameters parallel chart for 29 sample points with input parameters.

DOE sample points has sent to the CFD solver to update the value of the output parameter. These values are used to generate the response surface by fitting it using multidimensional regression techniques. In the current study, the Non-Parametric Regression (NPR) has been chosen as meta-modeling (surrogate) technique to create the response surface to meet the highly nonlinear behavior of the relation between the output parameter and input parameters.

Once the response surface has been created the so called Goal Driven Optimization (GDO) is performed based on screening method. The screening method allows generating a new sample set with 10,000 sample points and sort its sample based on the objective that is set to seek the maximum flow uniformity at the exit plane. The design candidates chosen by the optimizer are then verified using the CFD solver [19].

5. RESULTS AND DISCUSSION

Fig. 8 shows the generated response surface and the influence of the input parameters on the exit plane flow uniformity considering the optimal design set as a reference point. It is indicated that Fig. 8(1) shows the effect of both *a* and *b* on the uniformity. It can be observed that the uniformity is affected by changing parameter (a) as shown in Fig. 8(2). But, decreasing the trend of uniformity is attributed to increasing the parameter (b) as shown in Fig. 8(3). It can be seen from Fig. 8(4) that the maximum uniformity is occurred at 42% of the contraction length of the geometry profile, (i.e. inflection point) compared with the recommended ratio of 40%.

There is no guarantee for unique solution to have just one geometry but, there are some limitations should be considered to implement the design method. The optimization technique used has the capability to design the geometry. The design method gives three shapes (as shown in Table 1), all of them satisfy the requirements. The chosen geometry that can be validated experimentally and achieve the objectives. Table 1 shows the chosen design candidates and its verifications. It can be seen that candidate (*C*) has been chosen as the final design.

Table 1. Optimization design candidates and its verifications.

Candidate points	a	b	c	Uni_{ex}	Minimum wall shear
Candidate (A) Verification (A)	0.088835	0.036637	0.325706	0.981661211 0.979834	0.003984247 0.00579331
Candidate (B) Verification (B)	0.087155	0.034527	0.34500	0.98129169 0.979716	0.004430081 0.00658563
Candidate (C) Verification (C)	0.066435	0.035758	0.414229	0.980838786 0.980373	0.010024012 0.00972408

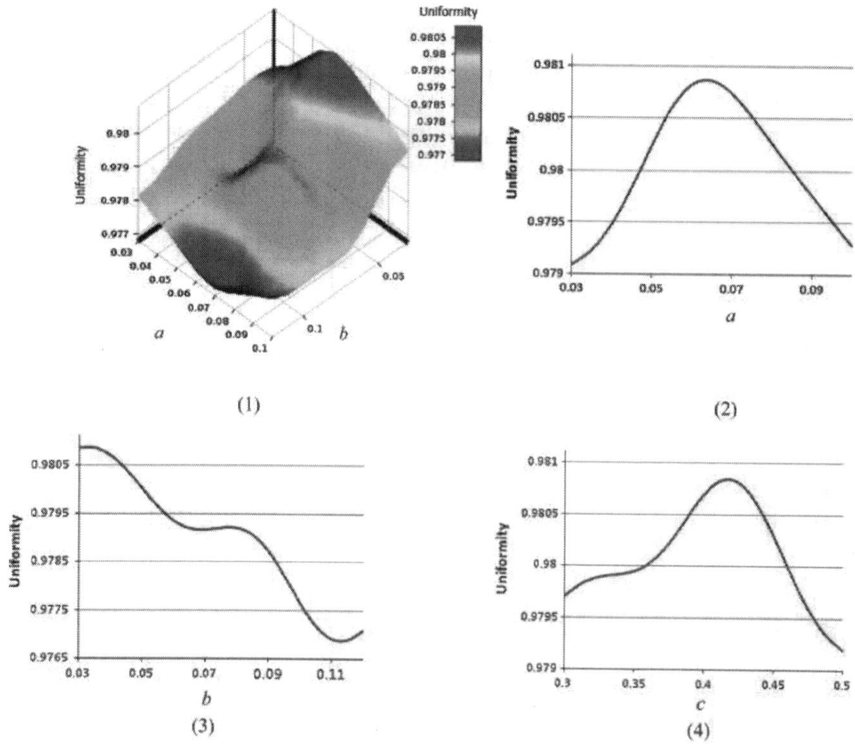

Figure 8. (1) Effect of both *a* and *b* on the uniformity, (2) effect of *a* on the uniformity, (3) effect of *b* on the uniformity and (4) effect of *c* on the uniformity.

The computational model was reviewed to test for uniformity of flow in the working section. Fig. 9 shows the velocity contour in the mid-plane of the contraction. It can be seen that the flow is fully developed at the test section. Also, it can be observed that the contraction is free from separation and no tracing of Gortler vortices. However, the relation between the radius of the concave shape of contraction and the momentum thickness is not comparable, where the radius = 1 m, and the momentum thickness $\theta = 0.003567$ m computationally while $\theta = 0.002338749$ m experimentally. Also, the optimization method with certain constrain prevents adverse pressure gradient that causes separation in the contraction and non-uniformities that may transfer to the working section as shown in Figure 9 and Figure 10, respectively. In addition that there is a positive minimum shear stress at the wall of contraction as shown in Fig. 11.

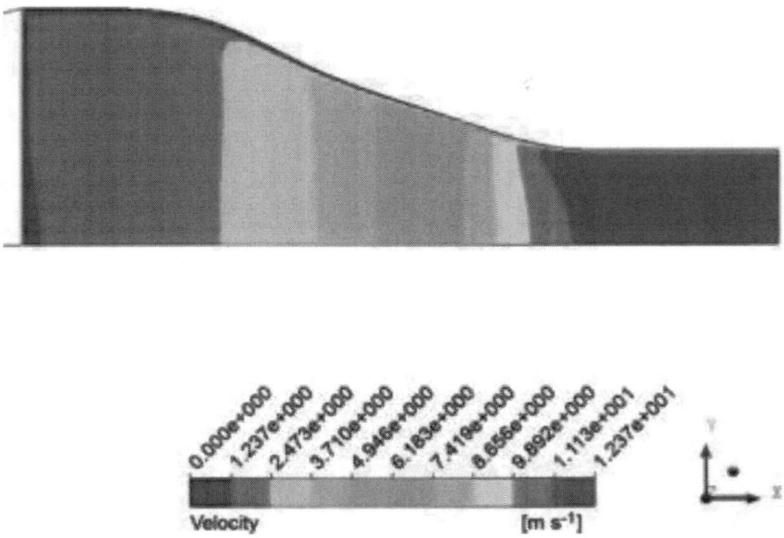

Figure 9. The velocity contour in the mid-plane of contraction.

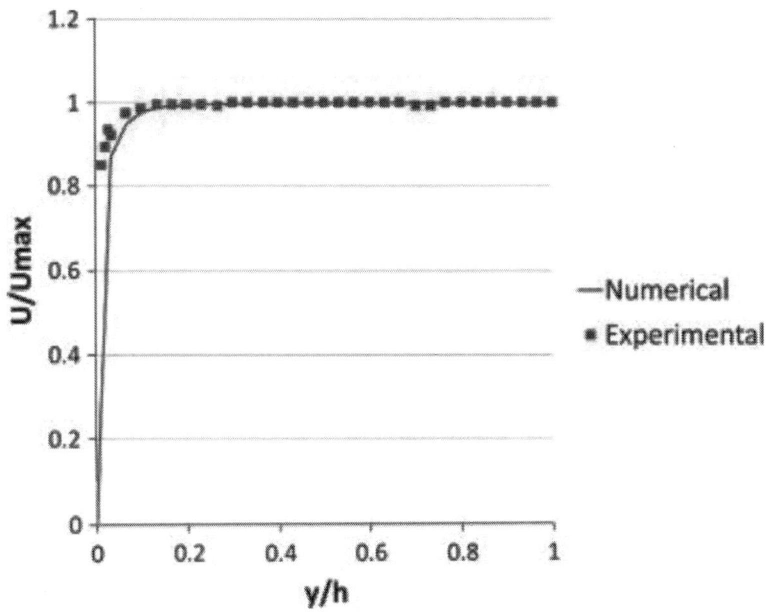

Figure 10. The velocity profiles at the mid exit plane.

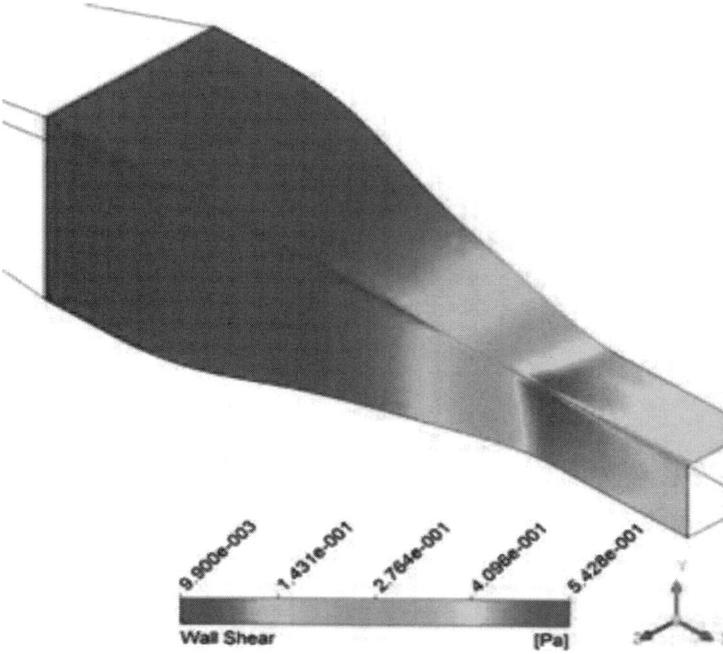

Figure 11. The contour of the wall shear of the contraction.

The computational model demonstrated reasonably uniform flow across the wind tunnel mid plane that appeared to have the most uniform velocity profile, and hence it was selected for manufacture. The velocity profiles (both of numerical and measurements) at this plane are shown in Fig. 10. Comparison of the uniform flow through the working section, between predicted and experimental results shows close agreement. However, it is noticed that there is small discrepancy (of order ±2%.) between the predicted results and measurements.

It was found that of the model tested experienced no separation. In Fig. 11, a typical wall shear plot for model demonstrates the lack of separating flow (indicated by τ_{min}, minimum positive values of wall shear over the entire wall).

The boundary layer characteristics at the exit of the test section are presented in Table 2. The computed boundary layer characteristics from the new design contraction show better agreement with results from experimental wind tunnel tests than the old design based on the analytical technique. It is noticed the improvement on the performance of the wind tunnel as the new design has 38.68% decrease in boundary layer thickness and 28.43% decrease in momentum thickness compared with the old design.

Table 2. Boundary layer characteristics at the exit of test section.

	Old	New	Experimental
Half hydraulic diameter $D_h/2$ (m)	0.4	0.3	0.3
Maximum velocity (m/s)	6.82286	12.1956	12.11892061
Dimensionless boundary layer thickness $2\delta/D_h$	0.181188	0.148143	0.11128
Boundary layer thickness δ (m)	0.072475	0.044443	0.033384
Displacement thickness δ^* using numerical integration (Trapezoidal method) (m)	0.009713	0.007492	0.002911251
The momentum thickness θ using numerical integration (Trapezoidal method) (m)	0.004984	0.003567	0.002338749
Reynolds number based on momentum thickness Re_θ	2168.764	2774.47	1807.596359

6. CONCLUSIONS

In the present contribution, a detailed parametric study is concerned with design optimization technique with (RANS) analysis of flow for design shape of three dimensional geometry of existing open circuit wind tunnel contraction. The present study achieves the recommended contraction ratio, maximum uniformity at the working section mid-plane, prevention of separation, without Gortler vortices in the contraction, and minimizing the boundary layer thickness at entrance to the working section.

The use of CFD has increased the flexibility of shape considered, and allowed the use of design exploration algorithm to optimize the profile of the contraction in an automated manner. The different design points were rated based on a goal-driven optimization using screening method to choose the best design set and verified by the CFD solver to define the profile.

It was found that the best result, producing the most uniform velocity profile at inlet to the working section, and preventing separation of the flow within the contraction, was obtained when the point of inflection was located as far downstream as possible.

The extensive numerical simulation illustrates that the hexahedral mesh is more stable and robust to converge. In spite of good results, usage of this type makes the mesh not fine enough near the wall to resolve the characteristics of the boundary layer.

The boundary layer characteristics based on optimization technique show significant improvement on the performance of the wind tunnel.

Physical calibration of the facility has validated the CFD method used and demonstrated that the technique can be addressed for future wind tunnel designs.

The entire study took less than a week compared to few weeks which would be required in the absence of well integrated optimization and CFD tools.

REFERENCES

1. Fang F, Chen JC. Experimental and analytical evaluation of flow in a square-to-square wind tunnel contraction. J Wind Eng Indust Aerodyn 2001;89:247–62.
2. Chmielewski GE. Boundary layer considerations in the design of aerodynamic contractions. vol. 11, no.8. St. Louis (MO): McDonnell Douglas Research Laboratories; 1974.
3. Dehghan Manshadi M, Mirzaei M, Soltani MR, Ghorbanian K. Control of pressure gradient in the contraction of a wind tunnel. World Acad Sci Eng Technol 2008;2:04–29.
4. Lanspeary PV, Bull MK. A mechanism for laminar threedimensional separation in duct contractions. In: 13th Australasian fluid mechanics conference; 1998.
5. Mehta RD. Aspects of the design and performance of blower tunnel components. Ph.D. Thesis. Department of Aeronautics, Imperial College, University of London; November 1978.
6. Bell JH, Mehta RD. Contraction design for small low-speed wind tunnels. JIAA report TR-64. Department of Aeronautics and Astronautics, Stanford University, NASA CR-177488; August 1988.
7. Bell JH, Mehta RD. Boundary layer predictions for small lowspeed contractions. AIAA J 1989;27(March):372–4.

8. Kim WJ, Patel VC. Origin and decay of longitudinal vortices in developing flow in a curved rectangular duct. J Fluids Eng 1994; 116:45–52.

9. Johl G, Passmore MA, Render PM. Design methodology and performance of an in draft wind tunnel. Aeronautical J 2004; 108(1087):465–73.

10. Bell J, Mehta R. Contraction design for small low speed wind tunnels. Technical report NASA-CR-177488, NASA; 1988.

11. Sargison JE, Walker GE, Rossi R. Design and calibration of a wind tunnel with a two dimensional contraction. In: 5th Australasian fluid mechanics conference. University of Sydney; 2004.

12. Morgans RC, Doolan CJ, Stephens DW. Derivative free global optimization of CFD simulations. In: 16th Australasian fluid mechanics conference. Crown Plaza, Gold Coast, Australia; 2007.

13. Hammam M. Investigation of the Performance of horizontal axis wind turbines equipped with gurney flaps. M.Sc. Thesis. Suez Canal University; 2009.

14. Mehta RD, Bradshaw P. Design rules for small low speed wind tunnels. Aeronautical J 1979;83(827).

15. Con JD, Rick CM. Numerical evaluation and optimization of low speed wind tunnel contractions. American Institute of Aeronautics and Astronautics; 2005.

16. Laws EM. The generation of velocity profiles in internal flows. Internal Report. University of Salford; 1971.

17. Eckert B, Pfluger F. The resistance of commercial round grids NACA T.M. 1003; 1941.

18. Design Xplorer Documentation. ANSYS Inc.; 2010.

19. ANSYS FLUENT User's Guide. ANSYS Inc.; 2010.

3
CHAPTER

Portable Wind Tunnels for Field Testing of Soils and Natural Surfaces

R. Scott Van Pelt[1] and Ted M. Zobeck[2]

[1] United States Department of Agriculture – Agricultural Research Service (USDA-ARS), Big Spring, Texas, USA

[2] USDA-ARS, Lubbock, Texas, USA

1. INTRODUCTION

Wind erosion of soils refers to the detachment, transport, and subsequent deposition of sediment or surface soils by wind. This process is sometimes termed aeolian movement and is responsible for the formation and migration of dunes, soil degradation in agricultural areas, and formation of deep loess deposits in areas downwind from major sediment sources. From cross-bedding in ancient sandstones, it has been determined that aeolian movement of soils and sediments has been occurring for eons and is a natural geomorphic process. Wind erosion affects over 500 million ha of land worldwide and is responsible for emitting between 500 and 5000 Tg of fugitive dust into the atmosphere annually [1]. These fugitive dust emissions contain a disproportional amount of soil organic carbon and plant nutrients and the winnowing and loss of these materials degrades the soil [2, 3].

Much of what we know about aeolian processes comes from wind tunnel-based investigations. The seminal work of Ralph Bagnold was largely conducted in a stationary suction-type wind tunnel 9 m in length [4]. Wind tunnels allow control over the wind and surface factors controlling aeolian movement and thus much more definitive investigations can be conducted in a shorter period of time than in the natural environment where these factors are highly variable in time and space. Other early aeolian researchers used wind tunnels to assess the erodibility of soil surfaces without plant residues based on the texture of the soil and relative abundance of aggregates too large to be entrained by the wind [5]. Large stationary wind tunnels have allowed sufficiently detailed understanding of the physical processes of aeolian movement that predictive models such as the Wind Erosion Equation [6] and the Wind Erosion Prediction System [7, 8] have been developed.

Stationary wind tunnels continue to be used for aeolian research at scales from single grain movement [9] through soil surface scale [10] to landscape scale [11]. The ability to control the humidity of the atmosphere has enabled scientists to study such sensitive processes as the electrostatic interactions between particles and electrical fields generated during aeolian activities [12]. Stationary wind tunnels have also been used to study abrasion effects of wind-driven sands on building materials [13], crop plants [14], bare crusted soil surfaces [15], and soil surfaces with microphytic crusts [16] as well as to compare and calibrate instrumentation for aeolian filed studies [17, 18].

Fugitive dust is perhaps the most visible product of aeolian activity and stationary wind tunnels have been used to study fugitive dust emissions from eroding soils. From wind tunnel testing of crusted soils and aggregates, it has been determined that sandblasting of these otherwise non-erodible features is responsible for much of the dust generated during aeolian events [19, 20]. Soluble salts such as $CaCO_3$ effects on dust emissions have also been investigated in stationary wind tunnels [21] as have complex and vegetated surfaces [22] and specific soils from Death Valley, a major dust source area in North America [23]. Although stationary wind tunnels have great utility, they are limited to testing disturbed soil surfaces that have been removed from their natural setting. The development of field portable wind tunnels has greatly expanded our ability to investigate aeolian processes in the field under controlled conditions.

2. PORTABLE WIND TUNNELS

Over the last six decades, portable wind tunnels have been developed and used on natural soil surfaces to measure the effects of soil surface characteristics and protective cover on soil erodibility and dust emissions [24]. In their simplest form, portable field wind tunnels must have at least three components: 1.) a self contained or at least portable power source such as an internal combustion engine, 2.) a fan or blower to induce air movement and create an artificial wind, and 3.) a working section that trains the wind from the blower over a finite area of soil surface. Portable wind tunnels in which the fan or blower pushes air through the working section are called pusher-type tunnels and if the fan or blower pulls the air through the working section they are called suction-type wind tunnels. Other components may include transition sections between the blower and the working section including a flow conditioning section and instrumentation to measure the wind speed in the working section and/or to capture sediment at the mouth of the working section. A typical portable field wind tunnel is presented in Figure 1.

The use of portable field wind tunnels has been traced back as far as the early 1940s, but the designers and builders did not publish retrievable documentation of their efforts. Austin Zingg, a mechanical engineer with the US Department of Agriculture, was the first to document the design and construction of a portable wind tunnel [25]. This wind tunnel was used to test the erodibility of crop field surfaces [26] and to assess the effects of roughness and drag based on pressure differentials across the soil surface tested [27]. Other early researchers built a portable wind tunnel to test the susceptibility of field-grown crops to abrasion from saltating particles [28]. A small suction-type tunnel was successfully used to test the threshold wind velocity necessary for particle movement on natural surfaces compared with disturbed surfaces and sieved soil [29]. Another very small suction-type portable wind tunnel has been used in Australia to determine the relative dust emission rates for a range of iron ores and road surfaces [30].

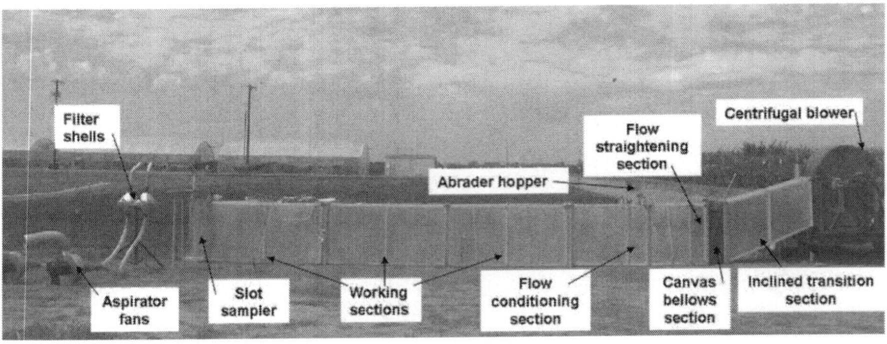

FIGURE 1. A Typical portable field wind tunnel showing component parts and sampling devices

Australians have also built a truck-mounted portable wind tunnel, tested rectangular and triangular working sections, and determined that the rectangular cross section was superior to the triangular one [31]. These same researchers noted the importance of wind flow conditioning upstream of the working section. Their wind tunnel has been used to assess the erodibility of bare cultivated and uncultivated soil [32], the effects of disturbance on the erodibility of cryptogamic crusts [33], and the sandblast injury and subsequent growth of narrow-leaf lupine [34].

In North America, a pusher-type wind tunnel was built to test the effects of oriented and random surface roughness elements on soil erodibility [35, 36]. This wind tunnel needed a small tractor and a secondary transmission for its power source and was transported using a large truck and 16 m long trailer. Another large portable wind tunnel built in North America was a suction-type wind tunnel that had a 12 m long working section. This wind tunnel was used to determine the erodibilities of natural crusted surfaces in North America and Africa [37-40]. A pusher-type wind tunnel with the power source and blower mounted on a truck bed and the working section lifted from the truck bed and lowered into place on the soil surface by hydraulic arms has been successfully employed to assess dust emissions from loess soils with and without surface cover in the Pacific Northwest of North America [24, 41-44] (Figure 2).

FIGURE 2. A large wind tunnel working section being lowered into place by a hydraulic arm.

Although large portable wind tunnels requiring mechanical devices to install may be powerful and allow testing of relatively large surface areas, the logistics of transporting them and finding a suitable footprint of level ground to test limit their utility. Examples of medium-size tunnels that may be installed by human power include a German tunnel that was field calibrated [45], a portable boundary layer wind tunnel with a working section formed of three 2 m long elements that fits on a 5 m trailer [46], and another German design that incorporates a rainfall simulator to induce wind-driven rain splash [47]. A summary of portable field wind tunnels, the dimensions of their working sections, maximum wind velocities developed, and reported boundary layer depths is presented in Table 1.

TABLE 1. Summary of previous and present portable wind tunnel designs, dimensions, maximum wind speed reported, and boundary layer thickness.

Publication	Tunnel Design	Width (m)	Height (m)	Length (m)	U_{max} ($m\ s^{-1}$)	Bdy. Lyr. (m)
Zing [25]	Pusher	0.91	0.91	9.12	17	0.23
Armbrust and Box [28]	Pusher	0.91	1.22	7.32	18	----
Gillette [29]	Suction	0.15	0.15	3.01	7	----
Fryrear [34, 35]	Pusher	0.60	0.90	7.00	20	0.15
Nickling and Gillies [37]	Suction	1.00	0.75	11.90	15	>0.2
Raupach and Leys [31]	Pusher	1.20	0.90	4.20	14	0.40
Pietersma et al. [24]	Pusher	1.00	1.20	5.60	"/>20	>1.0
Leys et al. [30]	Suction	0.05	0.10	1.00	19	----
Maurer et al. [45]	Suction	0.60	0.70	9.40	15	----
Van Pelt et al. [46]	Pusher	0.50	1.00	6.00	19	0.50
Fister and Ries [47]	Pusher	0.70	0.70	3.00	8	0.2

3. WIND TUNNEL DESIGN

In an engineering paper on wind tunnel design [48], the author stated that "the design of blower-driven air tunnel...is a combination of art, science, and common sense, the last being the most essential. It is difficult and unwise to predict firm rules for tunnel design." In addition to the power source, fan or blower, flow conditioning section, and working section, flow tripping fences and spires are often used to deepen the boundary layer thickness [49, 50], abrader feeders and regulators are used to initiate a saltation cloud, and sediment samplers to quantify the rate of erosion and dust emissions are often included in the design. Instrumentation such as anemometers of many designs and particle impact sensors are often used to monitor wind tunnel performance and to set operating parameters. In almost all cases, portable field wind tunnel designs are somewhat unique and highly influenced by their intended use.

3.1. Practical Design Criteria

When Zingg published the design and operation of his first portable field wind tunnel [25], he offered seven practical criteria to consider. These practical criteria are listed below;

1. The wind tunnel must be capable of producing an air stream free of general rotation and of known and steady characteristics.
2. It must provide easy and positive control of a range of wind velocities and forces common to the natural wind.
3. It must be durable.
4. It must be safe to use.
5. It should have sufficient size to afford free movement and representative sampling of eroding materials over field surfaces.
6. It must have ready portability.
7. It should be light in weight and amenable to quick and positive assemblage and dismantling.

Another criterion that he used but did not list was the use of commercially available equipment when available.

3.2. Aerodynamic Design Criteria

Mike Raupach and John Leys [31] suggested six aerodynamic criteria that should be considered in addition to the seven practical criteria proposed by Zingg. These aerodynamic criteria are listed below:

1. The flow must reproduce the logarithmic wind speed profile in the natural atmosphere, thus ensuring realistic aerodynamic forces on saltating grains.
2. The surface shear stress must scale correctly with the wind speed above the surface so that realistic aerodynamic forces act on grains of all sizes at the surface.
3. The vertical turbulence intensity and scale in the region close to the ground must be realistic, ensuring that vertical

turbulent dispersion of suspended grains is properly modeled.

4. The flow must be spatially uniform to avoid local scouring by anomalous regions of high surface stress.

5. Gusts should be simulated in the tunnel due to the fact that higher shear stress is required to initiate erosion than to sustain it.

6. A portable wind tunnel simulation of erosion should allow for the introduction of saltating grains at the beginning of the working section if more than the very upwind area of an eroding field is to be simulated.

They noted that criteria 1 to 4 are satisfied if the air flow near the ground surface is a well developed equilibrium boundary layer sufficiently deep to contain particle motion in the inner region where the mean wind speed profile is logarithmic and uniform over the eroding area. The logarithmic wind speed profile for neutral atmospheric stability has been described by:

$$U_z = (u^*/k)\, ln(z/z_o) \tag{1}$$

where U_z is the wind speed at height z above the surface, u^* is the friction velocity, z_o is the aerodynamic roughness length of the underlying surface, and k is the von Karman constant, usually assigned a value of approximately 0.4.

Criterion 5 requires turbulence with length scales greater than possible within the practical dimensions of portable wind tunnels and cannot be naturally generated by shear forces within either the working sections or flow conditioning sections of a portable wind tunnel. They tried to simulate gustiness using mechanical interruption of air flow in the flow conditioning section of their tunnel but discovered that the turning vane they employed for this purpose reduced the mean wind speed without increasing the vertical turbulence.

3.3. Simulating Saltation

Although criterion 6 is not truly aerodynamic, it is very necessary in order to simulate well developed steady state saltation of sand grains over an eroding surface. However, it also raises more questions as to the design and operation of the portable wind tunnel such as how much material to introduce, what the size distribution should be, and how to distribute it realistically in the flow before it strikes the ground surface tested in the working section. An orifice controlled gravity fed saltation initiator that drops the sand abrader into inclined tubes for acceleration before striking a sandpaper surface and bouncing into the flow stream is shown in Figure 3.

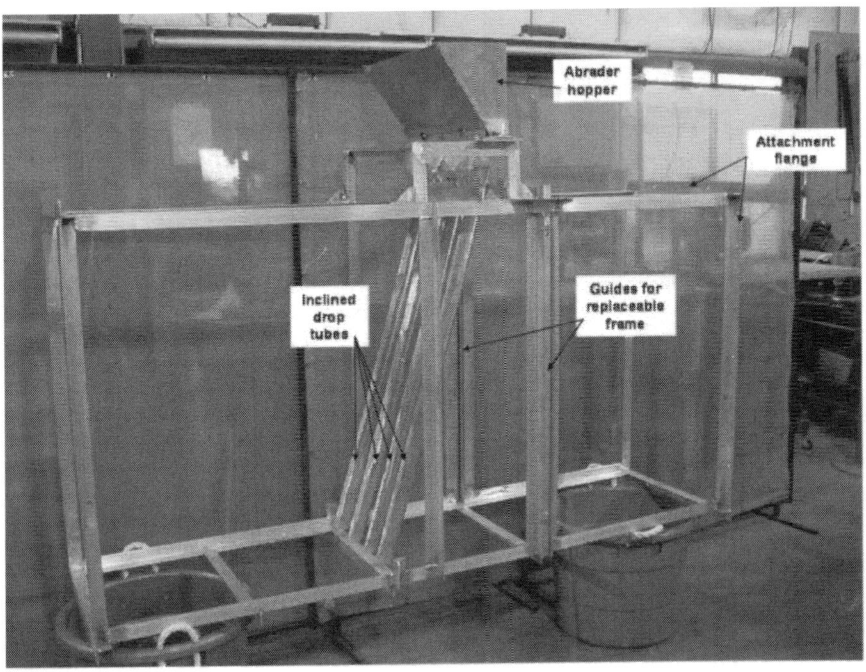

FIGURE 3. A complex flow conditioning section showing the abrader hopper and inclined tubes used to initiate saltation into the flow stream.

Saltation has been shown to reach a maximum at about 7 m length in wind tunnels [51] and decreases at longer distances, reaching equilibrium at between 10 and 15 m into the working section [45].

Longer working sections have limited utility however due to their lower transportability [47] and require a substantially longer uniform level surface on which to be set [52]. Working section lengths of portable field wind tunnels have varied from 3 m [19, 47] to almost 12 m [38, 39]. Recently, a small circular device named the Portable In-Situ Wind Erosion Research Laboratory (PI-SWERL) [52] has been used to develop shear stress over a surface and entrain particles using radially induced rather than linearly induced shear stress.

3.4. Power Sources

Power sources have ranged from external sources such as the power take-off shaft of a tractor as input to a transmission that output to drive chains [35, 36], to self-contained direct drive internal combustion engines [24, 25, 28, 31, 38-40], self contained internal combustion engines driving hydraulic pumps to provide for a hydraulic drive motor at the blower [46], and electric motors supplied by portable generators [45, 47]. All these power sources are field tested and reliable. The wind speed may be adjusted by varying the engine or motor speed or by changing the pitch of the fan or blower blades.

3.5. Fans and Blowers

The fans and blowers employed for wind tunnels are of two primary types. Axial fans (Figure 4a) are composed of fixed or adjustable pitched blades arranged radially around the axis of rotation, which is often aligned with the axis of flow through the wind tunnel. Although axial fans are highly efficient at inducing flow, the flow tends to spiral and this problem must be addressed [53] if the flow conditions of Zingg's first criterion are to be met. Centrifugal blowers (Figure 4b) have fixed pitch blades or impellers that are arranged parallel to the axis of rotation at the circumference of a blower cage. The axis rotation is commonly normal to the axis of air flow down the wind tunnel. Centrifugal blowers tend to be more flexible with respect to design, are more stable and efficient over a variety of flows, and produce less spiraling in the flow than axial fans [53].

Some portable field wind tunnels are too compact for adequate flow conditioning. This shortcoming is very problematic as flow considerations are the most important factor in the successful operation of the wind tunnel [31]. Wind tunnels may not reach true transport capacity or overshoot true transport capacity if flow conditioning upwind of the working section is inadequate [54] and wind tunnel height may limit the amount of upward mixing during strong turbulent diffusion [23]. The height of the working section affects the depth of the boundary layer that may be achieved. Upper limits of the Froude number F have been proposed for wind tunnel design of from 10 [55] to 20 [24]. The Froude number is defined by:

$$F = U^2 / gH \tag{2}$$

Where U is the wind tunnel design wind speed, g is the acceleration due to gravity, and H is the wind tunnel height. A well developed boundary layer at least 50 cm thick is recommended to ensure initiation of vertical particle uplift [45]. For this reason, mini-tunnels and micro-tunnels may be too small to allow results that can be scaled up to field scales [56].

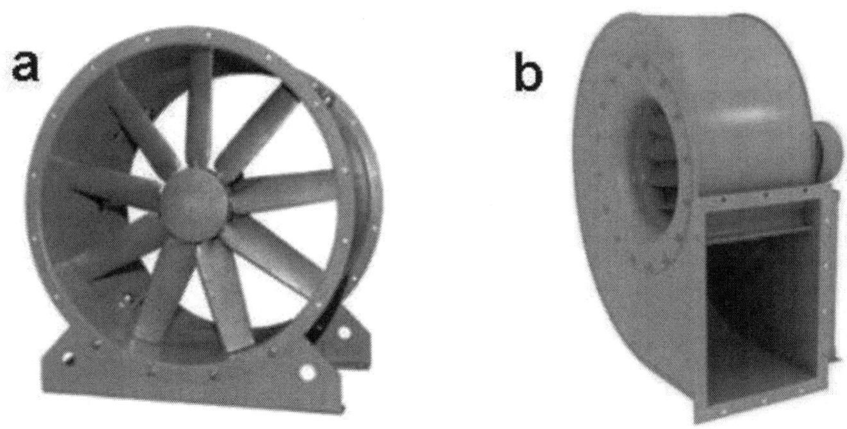

FIGURE 4. An axial fan (a) and a centrifugal blower (b) typical of those used in construction of portable field wind tunnels.

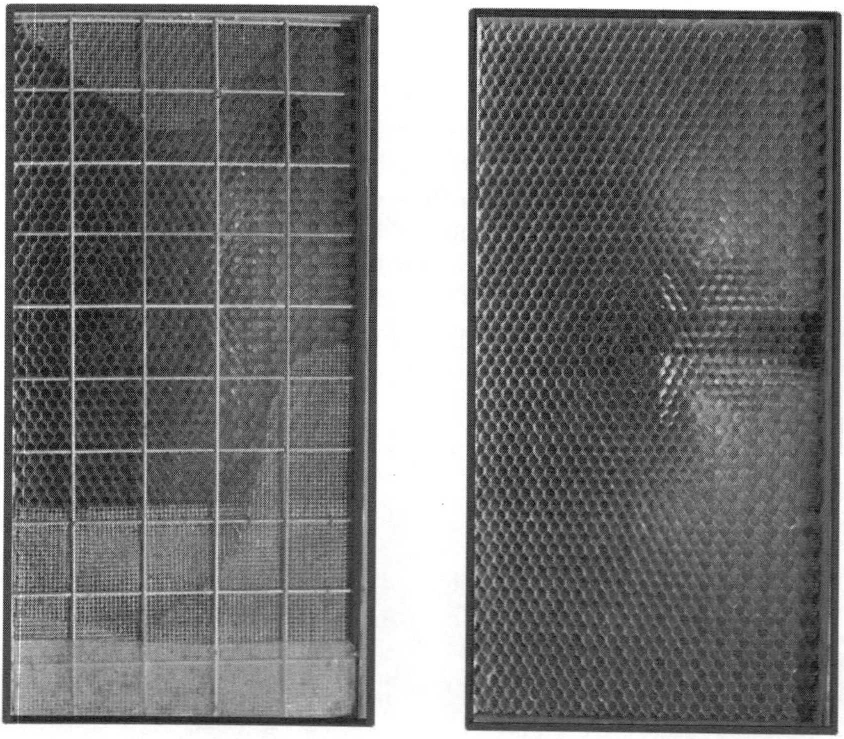

FIGURE 5. A flow conditioner showing the large cell honeycomb used to break the scale of eddies and straighten flow and also the 10 mm screen layers used to even the flow and create a logarithmic wind speed profile in the wind tunnel.

3.6. Flow Conditioning

Flow conditioning sections of various designs have been used to straighten the flow and remove or reduce the scale of eddies in the flow, to initiate a logarithmic wind speed profile and turbulence, and to initiate saltating abrader material into the air flow down the wind tunnels. A typical honeycomb flow straightener with 10 mm screen layers used to create an even logarithmic wind speed profile is presented in Figure 5. If the flow is properly conditioned and the height of the wind tunnel is not limiting the depth of the boundary layer may be estimated from the wind speed profile in the wind tunnel working section [46]. Investigators have stated that although boundary layer thickness is a poorly defined concept, it may be estimated as the height at which the wind speed profile attains 99 percent of its maximum value [57]. Finally, the proper regulation of carefully chosen abrader material allows for saltation clouds representing different rates of erosion and surface abrasion although rates consistent with those noted in the field for natural sand movement [58] are commonly used. Portable field wind tunnel may be used to estimate the threshold wind velocity necessary to initiate particle movement using impact sensors [18] or optically based sensors [59]. The technique of using the percentage of seconds in which moving particles are noted [60] is easily employed in a portable wind tunnel if the wind speed can be slowly and evenly increased.

4. CONCLUSIONS

Over the last 6 decades, portable field wind tunnels have been successfully used on several continents to study the controlling processes of aeolian particle movement, assess the erodibility of natural surfaces subjected to different disturbances, estimate dust emission rates for natural surfaces, investigate the partitioning of chemical and microbiological components of the soil on entrained sediment, and to estimate the threshold wind velocity necessary to initiate aeolian particle movement. Although not a perfect replacement for wind in the natural environment due to the absence of turbulent gusts, the forces created by the wind are repeatable and the accuracy of the tunnel is solely dependent on the accuracy of the devices measuring critical operating parameters such as wind velocity and sediment loading. When properly designed, calibrated, constructed, and operated, very useful information can be obtained in a relatively short period of time with these tools.

ACKNOWLEDGEMENTS

USDA is an equal opportunity provider and employer.

Nomenclature of symbols and their units

U– Wind velocity (m s^{-1})

z– Height above the surface (m)

z_0– Roughness length (m)

u* - Friction velocity (m s^{-1})

k– von Karman constant (~0.4)

F– Froude number (dimensionless)

g–Acceleration due to gravity (m s^{-2})

H– Wind tunnel height (m)

REFERENCES

1. Grini A, Myhre G, Zender C, Sundet J, Isakssen I. Model Simulations of Dust Source and Transport in the Global Troposphere: Effects of Soil Erodibility and Wind Speed Variability. Institute Report Series No. 124. Norway, University of Oslo, Department of Geosciences 2003.
2. Zobeck T, Fryrear D. Chemical and Physical Characteristics of Windblown Sediment: II. Chemical Characteristics and Total Soil and Nutrient Discharge. Transactions of the ASAE 1986; 29(4) 1037-1041.
3. Van Pelt R, Zobeck T. Chemical constituents of fugitive dust. Environmental Monitoring and Assessment 2007; 130 3-16.
4. Bagnold R. The Physics of Blown Sand and Desert Dunes. London: Methuen; 1941.
5. Chepil W. Properties of Soil Which Influence Wind Erosion: I. The Governing Principle of Surface Roughness. Soil Science 1950; 69(2) 149-162.
6. Woodruff N, Siddoway F. A Wind Erosion Equation. Soil Science Society of America Proceedings 1965; 29(5) 602-608.
7. Hagen L. Evaluation of the Wind Erosion Prediction System (WEPS) Erosion Submodel on Cropland Fields. Environmental Modeling and Software 2004; 19(2) 171-176.

8. Hagen L, Wagner L, Skidmore E. Analytical Solutions and Sensitivity Analyses for Sediment Transport in WEPS. Transactions of the ASAE 1999; 46(6) 1715-1721.

9. Huang N, Zheng X, Zhou Y, Van Pelt, R. Simulation of Wind Blown Sand Movement and Probability Density Function of Liftoff Velocities of Sand Particles. Journal of Geophysical Research 2006; D20201, doi:10.1029/2005JD006559.

10. Kohake D, Skidmore E, Hagen L. Wind Erodibility of Organic Soils. Soil Science Society of America Journal 2010; 74(1) 250-257.

11. Offer Z, Goossens D. Wind Tunnel Experiments and Field Measurements of Aeolian Dust Deposition on Conical Hills. Geomorphology 1995; 14(1) 43-56.

12. Zheng X, Huang N, Zhou Y. Laboratory Measurement of Electrification of Wind-Blown Sands and Simulation of Its Effect on Sand Saltation Movement. Journal of Geophysical Research 2003; 108(D10): 4322 doi:10.1029/2002/D002572.

13. Liu L, Gao S, Shi P, Li Y, Dong Z. Wind Tunnel Measurements of Adobe Abrasion by Blown Sand: Profile Characteristics in Relation to Wind Velocity and Sand Flux. Journal of Arid Environments 2003; 53(3) 351-363.

14. Baker J. Cotton Seedling Abrasion and Recovery by Wind-Blown Sand. Agronomy Journal 2007; 99(2) 556-561.

15. Zobeck T. Abrasion of crusted Soils: Influence of Abrader Flux and Soil Properties. Soil Science Society of America Journal 1991; 55(4) 1091-1097.

16. McKenna Neumann C, Maxwell C. A Wind Tunnel Study of the Resilience of Three Fungal Crusts to Particle Abrasion During Aeolian Transport. Catena 1999; 38(2) 151-173.

17. Goossens D, Offer Z. Wind Tunnel and Field Calibration of Six Aeolian Dust Samplers. Atmospheric Environment 2000; 34(7) 1043-1057.

18. Van Pelt R, Peters P, Visser S. Laboratory Wind Tunnel Testing of Three Commonly Used Saltation Impact Sensors. Aeolian Research 2009; 1(1-2) 55-62.

19. Gillette D. A Wind Tunnel Simulation of the Erosion of Soil: Effect of Soil Texture, Sandblasting, Wind Speed, and Soil Consolidation on Dust Production. Atmospheric Environment 1978; 12(8) 1735-1743.

20. Rice M, McEwan I. Crust Strength; A Wind Tunnel Study of the Effect of Impact by Saltating Particles on Cohesive Soil Surfaces. Earth Surface Processes and Landforms 2001; 26(7) 721-733.

21. Amante-Orozco A, Zobeck T. Clay and Carbonate Effect on Dust Emissions as Generated in a Wind Tunnel. In: Lee J, Zobeck T. (eds.) ICAR5/GCTE-SEN Joint Conference Proceedings 2002, Lubbock, TX, USA.

22. Kim D, Cho G, White B. A Wind-Tunnel Study of Atmospheric Boundary-Layer Flow over Vegetated Surfaces to Suppress PM10 Emissions on Owens (Dry) Lake. Boundary-Layer Meteorology 2000; 97(2) 309-329.

23. Roney J, White B. Estimating Fugitive Dust Emission Rates Using an Environmental Boundary Layer Wind Tunnel. Atmospheric Environment 2006; 40(40): 7668-7685.

24. Pietersma D, Stetler L, Saxton K. Design and Aerodynamics of a Portable Wind Tunnel for Soil Erosion and Fugitive Dust Research. Transactions ASAE 1996; 39(6) 2075-2083.

25. Zingg A. A Portable Wind Tunnel and Dust Collector Developed to Evaluate the Erodibility of Field Surfaces. Agronomy Journal 1951; 43(2) 189-191.

26. Zingg A. Evaluation of the Erodibility of Field Surfaces with a Portable Wind Tunnel. Soil Science Society of America Proceedings 1951; 15(1) 11-17.

27. Zingg A, Woodruff N. Calibration of a Portable Wind Tunnel for the Simple Determination of Roughness and Drag on Field Surfaces. Agronomy Journal 1951; 43(2) 191-193.

28. Armbrust D, Box J. Design and Operation of a Portable Soil-Blowing Wind Tunnel. USDA-ARS Pub. No. 41-131, US Govt. Print Off. Washington, D.C. 1967.

29. Gillette D. Tests with a Portable Wind Tunnel for Determining Wind Erosion Threshold Velocities. Atmospheric Environment 1978; 12(12) 2309-2313.

30. Leys J, Strong C, McTainsh G, Heidenreich S, Pitts O, French P. Relative Dust Emission Estimated from a Mini-Wind Tunnel. In: Lee J, Zobeck, T (eds.) ICAR5/GCTE-SEN Joint Conference Proceedings 2002, Lubbock, TX, USA.

31. Raupach M, Leys J. Aerodynamics of a Portable Wind Erosion tunnel for Measuring Soil Erodibility by Wind. Australian Journal of Soil Research 1990; 28(2) 177-191.

32. Leys J, Raupach, M. Soil Flux Measurements Using a Portable Wind Erosion Tunnel. Austalian Journal of Soil Research 1991; 29(4) 533-552.

33. Leys J, Eldridge D. Influence of Cryptogamic Crust Disturbance to Wind Erosion on Sand and Loam Rangeland Soils. Earth Surface Processes and Landforms 1998; 23(11).

34. Bennell J, Leys J, Cleugh H. Sandblasting Damage of Narrow-Leaf Lupine (Lupinus angustifolius L.): A Wind Tunnel Simulation. Australian Journal of Soil Research 2007; 45(2) 119-128.

35. Fryrear D. Soil Ridges-Clods and Wind Erosion. Transactions of the ASAE 1984; 27(2) 445-448.

36. Fryrear D. Soil Cover and Wind Erosion. Transactions of the ASAE 1985; 28(3) 781-784.

37. Nickling W, Gillies J. Emission of Fine-Grained Particulates from Desert Soils. In: Leinen M, Sarnnthein M. (eds) Paleoclimatology and Paleometeorology: Modern and Past Patterns of Global Atmospheric Transport, Series C: Mathematical and Physical Sciences, 282, Dordrecht, The Netherlands, Kluwer Academic; 1989. p133-165.

38. Houser C, Nickling W. The Emission and Vertical Flux of Particulate Matter <10 μm from a Disturbed Clay-Crusted Surface. Sedimentology 2001; 48(2) 255-267.

39. Houser C, Nickling W. The Factors Influencing the Abrasion Efficiency of Saltating Grains on a Clay-Crusted Playa. Earth Surface Processes and Landforms 2001; 26(5) 491-505.

40. Macpherson T, Nickling W, Gillies J. Dust Emissions from Undisturbed and Disturbed Supply-Limited Desert Surfaces. Journal of Geophysical Research 2008; 113: F02S04, doi:10.1029/2007JF000800.

41. Saxton K, Chandler D, Stetler L, Lamb B, Claiborne C, Lee B. Wind Erosion and Fugitive Dust Fluxes on Agricultural Lands in the Pacific Northwest. Transactions of the ASAE 2000; 43(3) 623-630.

42. Chandler D, Saxton K, Busacca A. Predicting Wind Erodibility of Loessial Soils in the Pacific Northwest by Particle Sizing. Arid Land Resource Management 2005; 19(1) 13-27.

43. Sharratt B. Instrumentation to Quantify Soil and PM10 Flux Using a Portable Wind Tunnel. In: Proceedings of the International Symposium on Air Quality and Waste Management for Agriculture. ASABE Paper No. 701P0907cd. St. Joseph, Michigan, USA; ASABE 2007.

44. Copeland N, Sharratt B, Wu J, Foltz R, Dooley J. A Wood-Strand Material for Wind Erosion Control: Effects on Sediment Loss, PM10 Vertical Flux, and PM10 Loss. Journal of Environmental Quality 2009; 38(1) 139-148.

45. Maurer T, Hermann L, Gaiser T, Mounkaila M, Stahr K. A Mobile Wind Tunnel for Wind Erosion Field Measurements. Journal of Arid Environments 2006; 66(2) 257-271.

46. Van Pelt R, Zobeck T, Baddock M, Cox J. Design, Construction, and Calibration of a Portable Boundary Layer Wind Tunnel for Field Use. Transactions of the ASAE 2010; 53(3) 1413-1422.

47. Fister W, Iserloh T, Ries J, Schmidt R. A Portable Wind and Rainfall Simulator for In Situ Soil Erosion measurements. Catena 2012; 91(1) 72-84.

48. Mehta R. The Aerodynamic Design of Blower Tunnels with Wide-Angle Diffusers. Progress in Aerospace Science 1977; 18 59-120.

49. Counihan J. An Improved Method of Simulating an Atmospheric Boundary Layer in a Wind Tunnel. Atmospheric Environment 1969; 3(2) 197-214.

50. Irwin H. The Design of Spires for Wind Simulation. Journal of Wind Engineering and Industrial Aerodynamics 1981; 7(3) 361-366.

51. Shao Y, Raupach M. The Overshoot and Equilibrium of Saltation. Journal of Geophysical Research 1992; 97(D18) 20559-20564.

52. Sweeney M, Etyemezian V, Macpherson T, Nickling W, Gillies J, Nicolich G, McDonald E. Comparison of PI-SWERL with Dust Emission Measurements from a Straight-Line Wind Tunnel. Journal of Geophysical Research 2008; 113: F01012, doi:10.1029/2007JF000830.

53. Mehta R, Bradshaw P. Design Rules for Small Low-Speed Wind Tunnels. Aeronautical Journal 1979; 83 443-449.

54. Hagen L. Assessment of Wind Erosion Parameters Using Wind Tunnels. In: Stott D, Mohtar R, Steinhardt G (eds.) Sustaining the Global Farm: Selected Papers form the 10th International Soil Conservation Organization Meeting, West Lafayette, IN, USA Purdue University and the USDA-ARS National Soil Erosion Laboratory 2001; p742-746.

55. White B, Mounla H. An Experimental Study of Froude Number Effect on Wind Tunnel Simulation. In: Barndorff-Nielsen O, Willets B (eds.) Aeolian Grain Transport, Volume I: Mechanics New York, New York, Springer-Verlag 1991; p145-157.

56. Fister W, Ries J. Wind Erosion in the Central Ebro Basin Under Changing Land Use Management: Field Experiments with a Portable Wind Tunnel. Journal of Arid Environments 2009; 73(11) 996-1004.

57. Schlichting H, Gertsen K. Boundary Layer Theory. Berlin, Heidelberg, New York, New York, Springer-Verlag 2000.

58. Namikas S. Field Measurement and Numerical Modeling of Aeolian Mass Flux Distribution on a Sandy Beach. Sedimentology 2003; 50(2) 303-326.

59. Sherman D, Bailiang L, Farrell E, Ellis J, Cox W, Maia L, Sousa P. Measuring Aeolian Saltation: A Comparison of Sensors. Journal of Coastal Research 2011; 10059 280-290.

60. Stout J. A Method for Establishing the Critical Threshold for Aeolian Transport in the Field. Earth Surface Processes and Landforms 2004; 29(10) 1195-1207.

4

CHAPTER

A Six Degrees of Freedom Dynamic Wire-Driven Traverse

Thomas J. Lambert, Bojan Vukasinovic * and Ari Glezer

Woodruff School of Mechanical Engineering, Georgia Institute of Technology, Atlanta, GA 30332-0405, USA

ABSTRACT

A novel support mechanism for a wind tunnel model is designed, built, and demonstrated on an aerodynamic platform undergoing dynamic maneuvers, tested with periodic motions up to 20 Hz. The platform is supported by a 6-DOF (six degrees of freedom) traverse that utilizes eight thin wires, each mounted to a servo motor with an in-line load cell to accurately monitor or control the platform motion and force responses. The system is designed such that simultaneous control of the servo motors effects motion within ±50 mm translations, ±15° pitch, ±9° yaw, and ±8° roll at lower frequencies. The traverse tracks a desired trajectory and resolves the induced forces on the platform at 1 kHz. The effected motion of the platform is measured at 0.6 kHz with a motion capture system, which utilizes six near-infrared (NIR) cameras for full spatial and temporal resolution of the platform motion, which is used for feedback control. The traverse allows different platform model geometries to be tested, and the present work demonstrates its capabilities on an axisymmetric bluff body. Programmable timed outputs are synchronized relative to the model motion and can be used for triggering external systems and processes. In the present study, particle image velocimetry (PIV) is used to characterize the realized wakes of the platform undergoing canonical motions that are effected by this new wind tunnel traverse.

Keywords: model support; 6-DOF; wind tunnel testing; aerodynamics; wakes

1. INTRODUCTION

An inherent difficulty with wind tunnel investigations of nominally "free" aerodynamic bodies is related to their mounting in the tunnel test section. Ideally, the model support should cause little or no aerodynamic interference (such as a magnetic-force support [1,2]), but most conventional support systems have relied on some form of a rigid connection that can interfere with the flow around the body and especially in its wake. The most common model support in wind tunnels is of a sting type (e.g., Achenbach [3]) that is typically connected to either aft or a side of the model. Another related support approach is by utilization of struts (e.g., Taylor *et al.* [4,5]), while the least common is a wire support (e.g., Griffin *et al.* [6]), which is usually restricted to low flow speeds. In a case where a model is desired to realize multiple degrees of freedom motion, a combination of sting and strut supports are often needed (e.g., Pattinson *et al.* [7]). Furthermore, there is a prevailing interest in model support interference with the "true" flow dynamics for transonic and supersonic flow regimes (see an extensive bibliography presented by Tuttle and Gloss [8]). This interference is primarily caused by the physical presence of a model support within the model wake, preventing its full development, and the shed vortices off the support structure that may interfere with the vortices naturally shed off the model. These problems have been recognized from early stages of wind tunnel testing (e.g., Bacon [9]), and different methods for either calibration or correction for the support interference have been proposed since about a hundred years ago.

As can be expected, support interference problems become exacerbated if the model support is to impose dynamic motion on the model. An extensive review of support interference in dynamic wind tunnel tests was presented by Ericsson and Reding [10], outlining a practical guideline for best practices in minimization of the support interference under specific flow regimes and model geometries. Beyers [11] pointed to the coupled interference of the support structure at the tunnel walls in the dynamic testing that precludes existence of any analytical corrections. He suggested that dynamic interference could be effectively eliminated in a case of rotary balance testing.

The current study extends the methodology for a thin wire support, aimed at minimizing flow interference with a traverse that can affect dynamic motion of a wind tunnel model at low flow speeds, and is primarily influenced by earlier studies of Abramson *et al.* [12,13] and Lambert *et al.* [14]. The present work investigates a wire-mounted body on a programmable six degrees of freedom (6-DOF) ($x/y/z$-translation and roll/pitch/yaw) eight-wire traverse that is electromechanically driven by a dedicated feedback controller to remove the parasitic mass and inertia of the dynamic support system and of the model. This traverse is designed to support any model with eight mounting points, aimed at estimating the aerodynamic force that the model without any support would experience, given: (1) The flow is not in the transitional speed range, where the wires' wake could alter the flow's transition to turbulence; and (2) The minimum length scale of the model is at least an order of magnitude larger than the diameter of the support wires to ensure decoupling of their respective vortex shedding.

For demonstration, an axisymmetric body is used for the purpose of illustrating the control authority of the current developed traverse system.

2. EXPERIMENTAL METHODOLOGIES

2.1. Physical Traverse Components

This traverse is designed as an add on to an open-return wind tunnel at Georgia Tech having a test section that measures 91 cm × 91 cm and a maximum speed of 40 m/s, and is also designed such that it can be incorporated and removed from the tunnel, if desired. The support mechanism uses multiple strands of wire rope (aircraft cables) to suspend an aerodynamic platform inside the wind tunnel, and is shown in Figure 1a. The wire ropes have a very small cross-sectional area of 1.13 mm^2 and they have low blockage, thus minimizing local acceleration of the airstreams in the vicinity of the model. Theoretically, only six wire ropes would be required for a statically-determinate support of the model. However, the geometry of the wind tunnel and the desire to keep the wire rope and actuators near the corners of the tunnel cross-section led towards an over-constrained support by eight wire ropes. The end of each wire rope is fastened to a servo actuator (*YASKAWA* SGMAV-10A, having peak torque of 9.55 N·m at a 6000 rpm peak), through a driving pulley. Each servo actuator is mounted outside of the tunnel and dynamically feeds a wire rope inward or outward to alter the position and orientation of the suspended aerodynamic platform. The corner flanges of the wind tunnel are modified to provide eight small openings with enough clearance to accommodate the desired range of wire motion and therefore model motion. It is noted that the motor locations are spaced vertically by 120 cm, horizontally by 128 cm, and streamwise by 176 cm, where these locations are chosen for convenience and can be optimized for future performance objectives, if necessary. All of the driving pulleys have a secondary attachment to an extension spring (with a constant of 208.4 N/m) that provides a preload tension on each wire when the model is in equilibrium (for the current case study, the preload is approximately 60 N). The tension in each mounting wire is measured using an inline miniature load cell (*FUTEK* FSH00097, with a range of 0–110 N, a non-repeatability of 0.055 N, and a mass of 0.02 kg). The temporal response of the raw load cell signals is on the order of 10 kHz, but these signals are pre-filtered and sampled at a rate of 1 kHz to match the controller rate (see Section 2.4). Figure 1b shows one constructed servo actuator assembly, with labeled components, as all eight servo actuator assemblies are identical.

Figure 1. Computer-aided design (CAD) model of the 6-DOF (six degrees of freedom) traverse (**a**); and a model of a servo motor assembly (**b**).

During the system design phase, several different types of motor actuators were considered for this application. Gear motor actuators could provide more output torque than the direct-drive configuration, but they would have too much inertia to meet the dynamic response goals. All types of linear actuators examined had limitations with response speed or maximum achievable wire rope tension, where the design goals for motion are 50 Hz controlled motions of model deflections of up to 5 mm, and wire tensions of up to 110 N. A vane-type hydraulic rotary actuator would be a compact, high torque, and high response rate solution, but it would require a hydraulic reservoir and pressurization system, and it was deemed less applicable compared to the realized rotary servo actuator shown in Figure 1b, although it would also achieve all of the design goals.

2.2. Motion Capture System

The primary goal of this traverse is to achieve accurate real time motion control of an aerodynamic platform with disturbance rejection. A *VICON* MX Motion Capture System was acquired for measuring the spatially- and temporally-resolved motion of the wire-mounted model in the wind tunnel (Figure 2). This system consists of: (1) six MX-T40S cameras, each with a 12.5 mm lens (capable of up to 515 fps at full frame, and 2000 fps with a limited view), an electronic freeze frame shutter, and a resolution of 4 megapixels; (2) a MX-Giganet interface of the cameras to a camera host computer (provides power, synchronization, and data transfer); and (3) *VICON* Tracker 2.0 Software for low latency data streaming to the traverse controller computer.

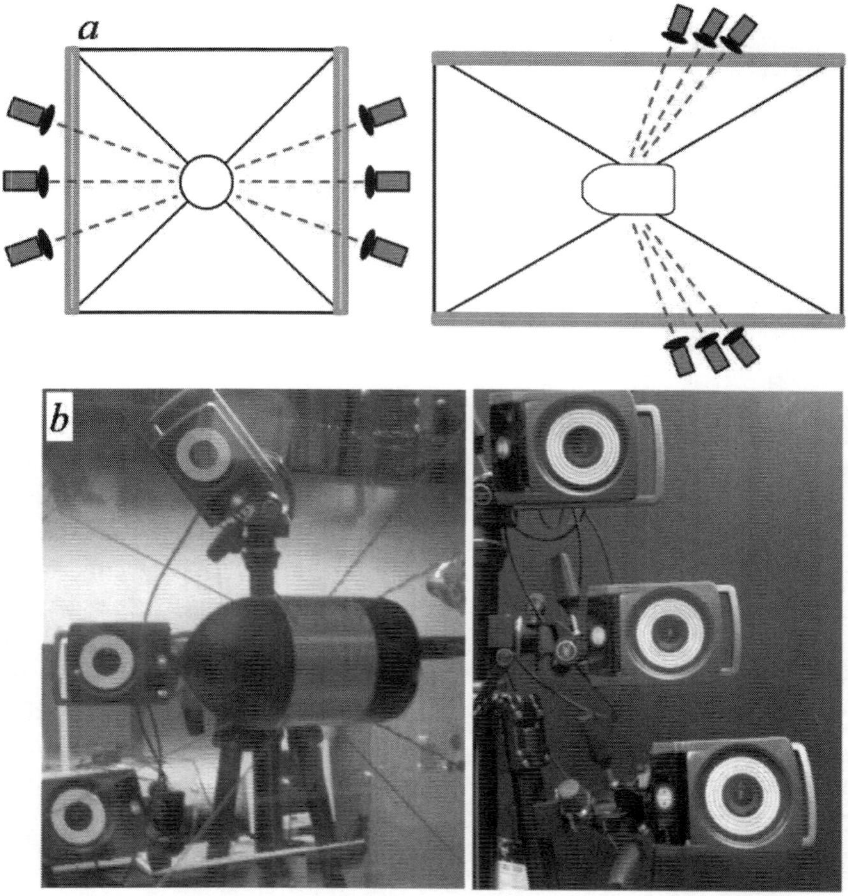

Figure 2. Schematics (**a**) and images (**b**) of the motion analysis system camera orientation.

The cameras are mounted around the wind tunnel's test section and are focused on the wire-mounted model. The current system utilizes overlapping fields of view to determine the time-resolved

(6-DOF) position of the moving model in the wind tunnel at a frequency of 600 Hz. The model's position is tracked using an optical correlation based on an array of small reflective round markers that are affixed to the model to reflect near infrared (NIR) light that is emitted by diode array built into each camera. The placement of the cameras can be arbitrary as long as the six projected images of these markers measured by the cameras contain enough information to fully define the 6-DOF motion of the model (shown in Figure 2a with both the front and top views of the system). The camera system is calibrated by dynamically moving a calibration target with three reflective markers through the measurement volume. Once the system is calibrated, a static image of a calibration plate of reflective markers is measured to construct a template that is saved on the host computer to transform the marker measurements into the dynamic aerodynamic model coordinates with respect to the wind tunnel frame of reference. An example of the chosen camera locations are shown in Figure 2b. After the system is calibrated, the *VICON*Tracker software calculates the error in each marker (having typical values between 0.05 and 0.5 mm, dependent on the calibration and camera locations), which are used to calculate the error in the measured model coordinates.

2.3. Case Study: Axisymmetric Model

This traversing system is designed to handle a variety of models including munitions, small-scale unmanned aerial vehicles (UAVs), parafoils, parachutes, and rotorcraft configurations. For the purposes of testing of the traverse, the model used as an example case study is an axisymmetric body, as shown in Figure 3, which has been of interest in prior investigations (e.g., [12,13,14]). This axisymmetric wind tunnel model (Figure 3) is assembled using stereo-lithographed (SLA) and aluminum components (D = 90 mm, L = 165 mm, m = 0.53 kg). The mid-section and nose are hollow and connected to the tail section by a central shaft. After assembly, the SLA model surface is painted black resulting in an average surface roughness of approximately 5 μm and then machined flush on a lathe with the central piece. In addition, six reflective circles of 7 mm diameter are attached on the surface for the motion analysis system, and each marker was made from 3M

7610 retroreflective tape of 0.4 mm thickness. These markers are arbitrarily placed so that each camera sees three of them (three of which can also be seen in Figure 2b and Figure 3c).

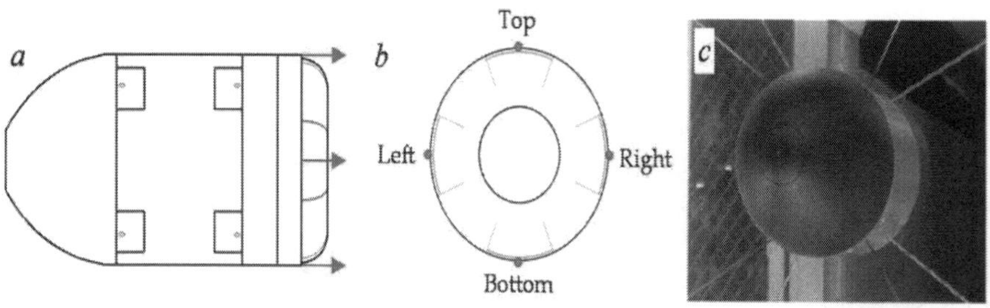

Figure 3. Side (**a**) and upstream (**b**) views of the CAD wind tunnel model with four hybrid fluidic actuators marked in green, and the model supported by the traverse (**c**).

A secondary requirement for this traverse is that it can provide electrical power to the wind tunnel model to enable future use of flow control applications. In the axisymmetric model example, four independently-driven synthetic jet actuators (labeled the top, left, bottom, and right actuators, as shown in Figure 3b) would be, in principle, integrated and distributed along the perimeter of the tail section, as they would comprise the flow control components in a follow-up study. The electrical power that would be used as a source for the synthetic jets is routed through weaved positive voltage lines (30 AWG) along the mounting wires on the downstream end of the model (shown in Figure 3c), where the entire system is grounded.

The range of translational and rotational motions of the traverse are heavily dependent on the location where each mounting wire is attached to the model (the range of translational motions are also dependent on the locations of the eight servo actuators). For example, if the wires are mounted such that they are equally distributed in the azimuth with 90° between each wire, a roll motion would be impossible, and this scenario is shown in Figure 4a. If the wires are not equally distributed in the azimuth, two symmetric groups of wires can be tightened or loosened to provide roll, as shown in red and green in Figure 4b. The theoretical maximum roll (with full controllability) occurs when one of these symmetric groups becomes in line with the center of the wire mounts, where this limit case is shown in Figure 4c, which occurs if the green set of wires in Figure 4b are tightened. A similar argument can be applied to both the pitch and yaw directions of the model looking at the side of the model or the top of the model, respectively. It is noted that in the front view shown in Figure 4, the projections of the four upstream and four downstream wires overlap, and these respective overlapping wires need to be tightened or loosened equally in order to prevent any pitch or yaw dynamics. The current mounting locations were chosen such that the model can theoretically have a significant and comparable range in all three rotational directions. These realized locations are centered on the model and displaced vertically by ±35.7 mm, horizontally by ±27.4 mm, and streamwise by ±26.3 mm. The extent of the realized traverse motion is discussed in detail in Section 3.1.

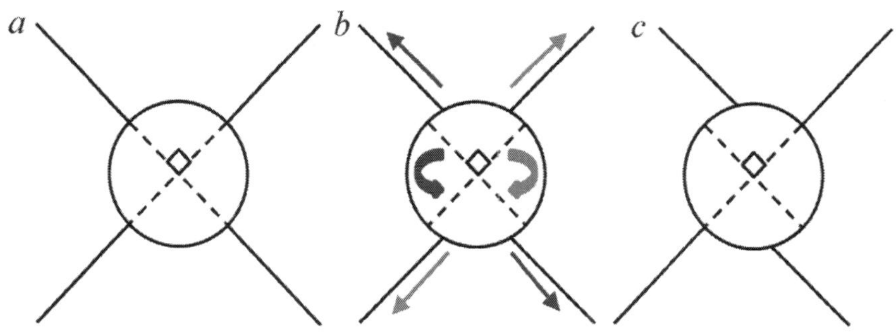

Figure 4. Orientation of the support where pure roll is disabled (**a**) and enabled for clockwise roll (green arrows) and counter-clockwise roll (red arrows) (**b**); along with its respective maximum clockwise roll deflection (**c**).

2.4. Electrical Traverse Components

The electrical subsystem of this traverse can be considered as being comprised of three separate composite elements. The first element is a "host" computer which is used for implementing the real time traverse controller and has two *Quanser* Q8 real time data acquisition boards which are used for signal communication of the servo motor command and encoder signals, as well as the flow control actuator commands. The second element is a high power distribution system that routes up to 1 kW power to servo amplifiers for each of the servo motors. Particular attention was given to proper wire routing and grounding in order to minimize electromagnetic interference (EMI) contamination of the measured signals. Each servo motor has an integrated 20 bit differential quadrature absolute encoder, where the output is converted to a single ended signal by custom electronics, and is sampled by the host computer through one of the Q8 boards. The third element is a custom multichannel Ethernet data acquisition (DAQ) system that samples the load cells. The load cell signals are on the order of mV and are susceptible to noise, so each load cell is routed with a short cable to signal conditioning electronics that have an instrumentation amplifier and passive resistor-capacitor (RC) filtering. These signals are then transmitted through universal data protocol (UDP) packets at a nominal frequency of 1 kHz to the host computer.

The user interfaces with a master computer that uses *Simulink* as an interface to build the controller. Upon completion of the desired controller, *Simulink* generates a "C" code that is cross compiled into a *QNX* binary. This binary is copied across an Ethernet network to the host computer (which utilizes a *QNX* operating system). When the master computer sends the execution command, the controller binary on the host machine is executed in a hard real time environment, which sends time-stamped signals back over Ethernet to the master computer for accurate data recording. The host machine is also connected via Ethernet to the broadcasted signal from the *VICON* machine to use the position measured from the camera system for feedback to the control system code. It should be noted that the driving signals for flow control can also be generated by the controller and output from the host computer for referencing/triggering of external waveforms and signals.

2.5. Control Systems Design

The designed traverse is controlled through the trajectory-tracking controller depicted in Figure 5. First, the user can provide two command inputs; a time trace for the desired model trajectory in 6-DOF, and a time trace for the actuation commands to, for example, amplitude modulate all four synthetic jet carrier waveforms, when flow control needs to be utilized. Second, the 6-DOF commanded motion is converted into eight servo motor commands (6-DOF to 8-DOF), which is calculated through the geometry of the model and chosen mounting points, assuming the wires are incompressible. The command signal to each servo actuator is then generated using eight proportional-integral-derivative (PID) controllers with the same coefficients ($\kappa_{P,inner}$ = 4.24 N·m/rad, $\kappa_{I,inner}$ = 21.20 N·m/(rad·s), ($\kappa_{D,inner}$ = 0.02 N·m·s/rad) which commands the torque in the motors. It is noted that with pure feedback control, there will be a phase lag between the desired command and the realized command, so the motions are currently restricted to time harmonic motions where optimal PID coefficients yield an output that is self-similar to the desired command with a change in phase and amplitude. The realized motion is then calculated through measuring the conditioned encoder signals and inverting these servo motor positions to the model position using a least squares algorithm (8-DOF to 6-DOF). This control loop is considered the "inner loop controller" in Figure 5. It is noted that this inner loop controller is sensitive to errors in the geometric parameters of the traverse (e.g., mounting locations, or damping and friction of the motors). In order to compensate for this geometric model, as well as the change in the amplitude of the desired command, an "outer loop controller" shown in Figure 5 is implemented to adjust the command of the inner loop based on the motion analysis system to allow for accurate trajectory tracking. The outer loop uses an integral error feedback controller that is set such that the measured trajectories reach the desired ones within 10 s ($\kappa_{I,outer}$ = 0.3 s^{-1} for the 3-DOF translational commands and 0.1 s^{-1} for the 3-DOF rotational commands). In addition, a manual gain of the inner loop command is implemented, determined by matching the magnitudes of the measured amplitude to the desired amplitude, which is dependent on the frequency and desired motion.

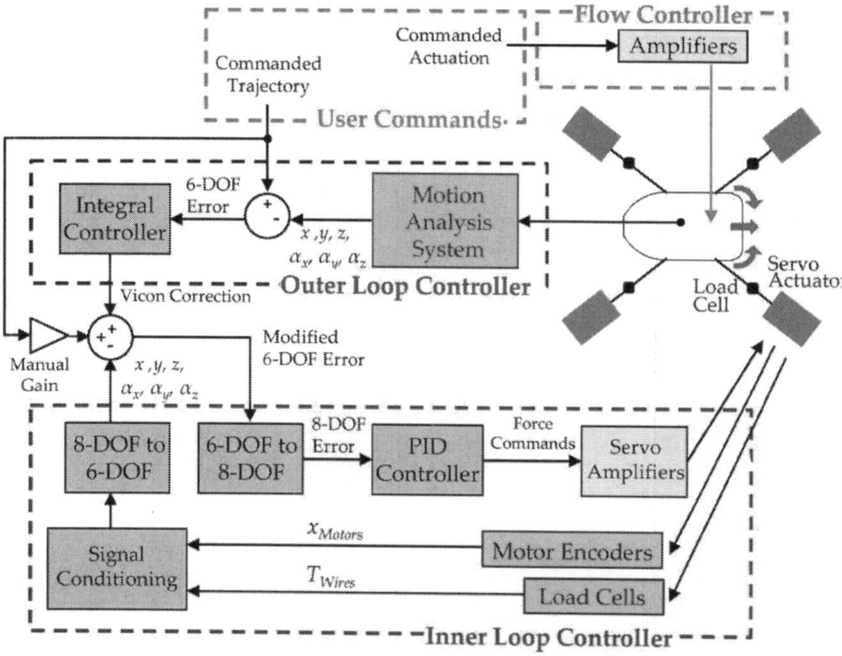

Figure 5. Schematics of the traverse trajectory-tracking controller.

The real time wire tensions measured through the load cells are used to extract the aerodynamic forces and moments on the model. This is realized by using a free body diagram in a control volume (CV) shown in Figure 6, where the aerodynamic forces are calculated by the inertial force, FI−→, of the model (acting at the center of gravity, CG, and calculated from its estimated acceleration) with all other forces subtracted. These other forces include the weight of the model (FG−→, acting at CG), the tension in the wires, FT−→, and the estimated aerodynamic drag of the wires alone, FA−→wires. The latter is calculated assuming the wires are cylinders with pure form drag (C_D = 1.25) and no friction drag, using the real time location of both the model and motor wire ends and calculating the normal projection of the flow. It is noted that each wire force is estimated as acting on the midpoint between the tunnel flange and the respective real time location of the wire mount on the model, and acting in the direction normal to the wires. The resultant force is the aerodynamic force on the model, FA−→model , and a similar approach can be used to

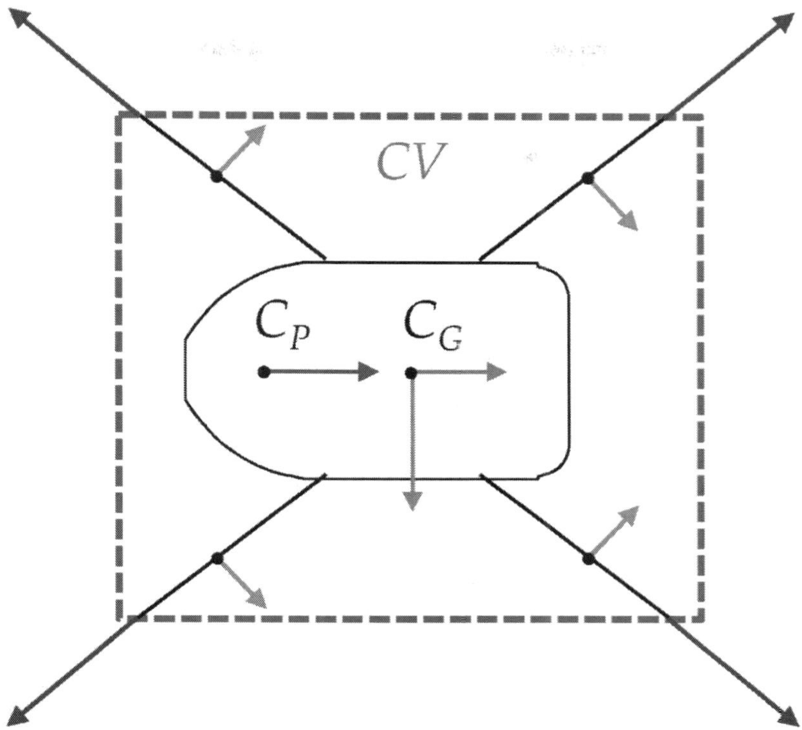

$$\overrightarrow{F_{A,model}} = \overrightarrow{F_I} - \overrightarrow{F_G} - \Sigma\overrightarrow{T} - \Sigma\overrightarrow{F_{A,wires}}$$

Figure 6. Free-body diagram of the aerodynamic platform and the support wires with the aerodynamic model (blue), inertia (green), gravitational (cyan), and aerodynamic wire (orange) forces, and the wire tensions (red).

determine the moments by weighting each force by the location at which it acts. Measurements of the aerodynamic forces for the axisymmetric body using this force balance are presented inSection 3.2. The current trajectory-tracking controller described in this section restricts the model to a prescribed trajectory, where the model has a negligible response to external aerodynamic forces (where the disturbance force is on the order of ~1 N, and the spring constant of the cables are on the order of ~100,000

N/m). It should be noted, however, that a free flight controller can instead be designed where the model would respond directly to the disturbance forces (by using the load cells as the primary feedback instead of the encoders and camera signals), but a more aggressive controller would have to be implemented in the inner loop to induce dynamics that do not significantly lag the induced forces, and this type of controller is considered for future work.

3. RESULTS

3.1. Motion Response

The system dynamic performance is characterized in terms of displacement range and frequency of sinusoidal commands, and is depicted in Figure 7. Although the mounting points of the wires on the motors and the model determine the ideal maximum range of motion with limitless tension and unbreakable wires, this range diminishes for a fixed maximum tension. This range is also frequency dependent and diminishes with higher frequencies as the wires have to support an accelerating model with a larger inertial force. Therefore, the current controller is tested with a range of frequencies and the maximum range of motion in all six degrees of freedom is measured with a maximum wire tension of 110 N, and the results are shown for translations and rotations, in Figure 7a,b, respectively. This frequency response is dependent on all of the aforementioned geometric and controls parameters and can be further optimized by modifying these parameters, if necessary. The measured displacement amplitude that is realized is 50 mm in translation (x, y, and z), and rotations of 8° αx, 15° αy, and 9° αz at a frequency, f = 0.5 Hz, and decreases to 5 mm (x, y, and z), 1° αx, 3° αy, and 2° αz at f = 20 Hz. Four of these motions are visualized in Figure 7c, with streamwise (x) translation, vertical (z) translation, pitch (αy), and yaw (αz).

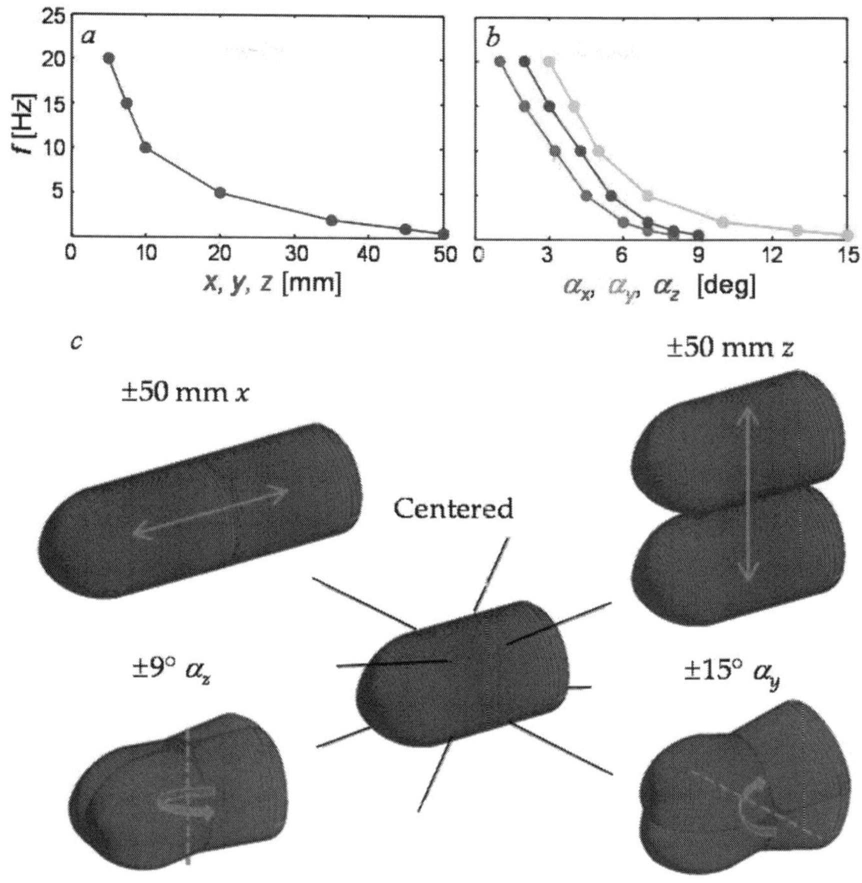

Figure 7. Frequency response of tested translational (**a**), and rotational (**b**) motions, with an illustration of the maximum traverse displacements achieved in four of the degrees of freedom (**c**).

To test the control authority of the traverse for 6-DOF motion tracking, a combination of rotations is commanded (pitch and yaw out of phase) and the instantaneous measurement is shown in Figure 8 with streamwise translation x (Figure 8a,b), cross stream translation y (Figure 8c,d), vertical translation z (Figure 8e,f), roll αx (Figure 8g,h), pitch αy (Figure 8i,j), and yaw αz (Figure 8k,l), respectively. The commanded signal is a 3°

amplitude sinusoidal pitch and sinusoidal yaw, 90° out of phase with each other. This same motion is executed with a representative "low" frequency of 1 Hz in Figure 8a,c,e,g,i,k along with a "high" frequency of 10 Hz in Figure 8b,d,f,h,j,l. Both of these motions are executed with wind tunnel flow (Re_D= 2.3 × 10^5) to demonstrate the disturbance rejection of the controller. The time resolved path captured by the motion analysis system is shown in blue while the commanded signal is shown in dotted gray. These data show significant agreement between the commanded and measured signals in terms of peak-peak magnitude (within 5% at 1 Hz in Figure 8i,k and 7% at 10 Hz in Figure 8j,l), with a phase lag of 40° at 1 Hz and 30° at 10 Hz. The realized deviations in the four other degrees of freedom that are not commanded are small: within 0.5° in roll (Figure 8g,h), and within 0.3 mm in all translations (Figure 8a–f). As mentioned in Section 2.2, the errors in the motion measurements for this data set (which is the same as for the motions in Section 3.2 and Section 3.3) are calculated using the motion analysis system calibration error output (0.45 mm per marker on average), yielding errors of 0.18 mm in translation and 0.02° in rotation. It is noted that these instantaneous traces are measured 10 s after the traverse is activated after the transient response of the traverse has settled. The measurements that are shown in Figure 8 can also be used as feedback to the triggering lines for flow diagnostics (e.g., particle image velocimetry (PIV)) and/or flow control, and therefore as long as the flow control is a function of the measured DOF the phase lag between the commanded and measured trajectory is inconsequential for the considered harmonic motions. In this section, the wind speed is fixed at 40 m/s and it is noted that the traverse controller treats the external aerodynamic forces on the model as disturbances to be rejected with PID control. Keeping fixed PID coefficients, the phase lag of the commanded and realized responses would increase with increasing wind speed, as the disturbance force grows, but in the present investigations, the PID parameters are pre-adjusted for a given Re_D and can be further tuned with a change in the wind speed.

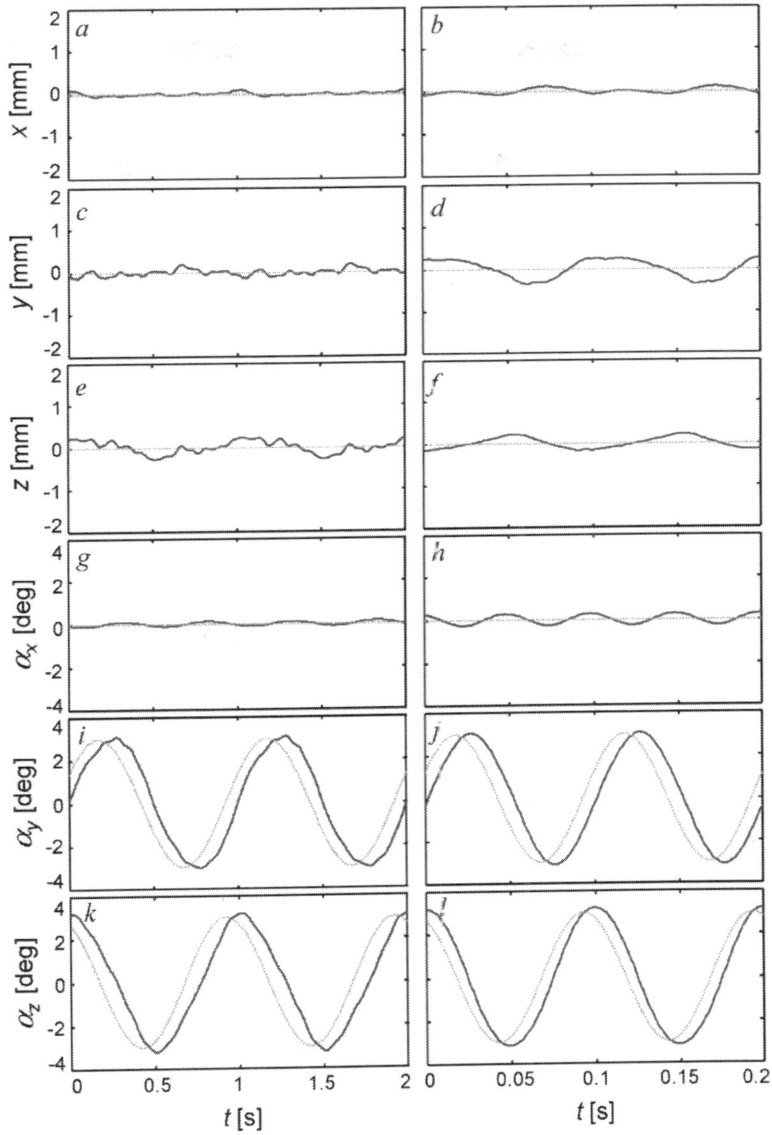

Figure 8. Traverse motion response for 3° amplitude pitch and yaw rotation, 90° out of phase, with measured translation in *x* (**a**,**b**), *y* (**c**,**d**), and *z* (**e**,**f**), and rotations in roll (**g**,**h**), pitch (**i**,**j**), and yaw (**k**,**l**), for frequencies of 1 Hz (**a**,**c**,**e**,**g**,**i**,**k**) and 10 Hz (**b**,**d**,**f**,**h**,**j**,**l**), with a flow $Re_D = 2.3 \times 10^5$. The commanded and measured responses are shown in dotted gray and blue, respectively.

3.2. Force Response

The second analysis that is implemented on the traverse is a validation of the force and moment measurements. Initially, the drag force on the stationary centered model (F_D) is measured over a range of wind tunnel speeds, U_0, of up to 40 m/s ($0 < Re_D < 2.3 \times 10^5$), and these data are presented in Figure 9a. The measured drag coefficient is $C_D = 0.239$, which is shown in a dotted line, and is in good agreement with the expected drag coefficient of $C_D = 0.25$ (for a similar ogive model with a sharp aft end, see Hoerner [15]). For this stationary centered model, all other forces are nominally zero, and are omitted for brevity. It is noted that for all of the forces presented in this section the force balance method shown in Figure 6 (see Section 2.5) is applied: the combination of the inertial and gravitational forces is measured first from the model executing the prescribed motion command without wind in the tunnel, and the aerodynamic force on each wire is calculated assuming the wires are cylinders with pure form drag ($C_D = 1.25$). The drag on each of these wires is normal to the wire and, in general motion, projects components in all six of the measured forces and moments in the system. For Figure 9, the gravitational force measured (model, wires, and load cells) is a constant, there is no inertial force, and the estimated aerodynamic forces on the wires add up to a pure drag force. These forces are subtracted off of the raw measured forces, which yields the aerodynamic drag force that is used to calculate the model C_D. When the model is set in motion, the projected gravitational moment changes with time as the location of the center of gravity varies, and the inertial force and aerodynamic forces on the wires also change with time. All of these time-varying forces are predicted (inertia and gravity effects are estimated from a commensurate measurement without wind) and subtracted from the raw measurement in order to yield the estimated aerodynamic forces. In addition, for the remainder of this paper the aerodynamic moments are measured relative to the center of the model wires at 0.54 L from the nose. It should be also noted that the wires do not only affect aerodynamic measurements, but also present additional sources of vorticity in the flow. However, increased levels of the flow fluctuations behind the wires do not appear to alter the dominant body wake dynamics, and the coupling between the two is deemed insignificant (see Lambert *et al.* [14]).

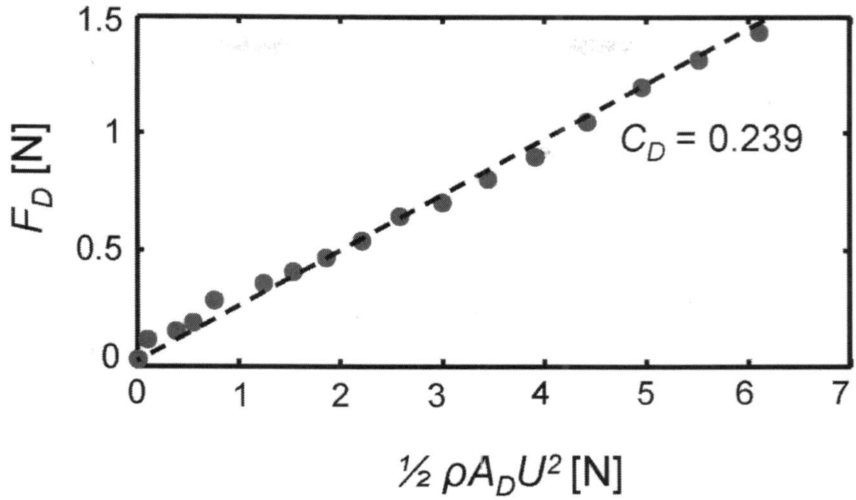

Figure 9. Measured axisymmetric model drag with varying wind tunnel speed.

Once the force on a centered stationary model is characterized, three canonical motions are investigated: dynamic pitch, plunge, and streamwise displacement. The first motion presented in Figure 10 is dynamic αy = ±3° (±0.052 radians) amplitude, simple harmonic pitching. This motion is quantified by the drag and lift force coefficients CD,L=8FD,L/ρπD2U20 as well as the pitching moment coefficient CD,L=8MP/ρπLD2U20, and for the rest of this force investigation, the free stream is set to U_0 = 40 m/s (Re_D = 2.3 × 10^5). For these flow conditions, the measured side force and roll and yaw moments are negligible in all three of these canonical motions, and they are omitted for brevity. Figure 10a–e show C_D, Figure 10f–j show C_L, and Figure 10k–o show C_M with the pitching angle, αy. Figure 10a,f,k show the a quasi-steady response at f = 1 Hz, (or a reduced frequency of k=πfDU0 = 0.013), and the increased frequency to 5 Hz (k = 0.065, Figure 10b,g,l), 10 Hz (k = 0.130, Figure 10c,h,m), 15 Hz (k = 0.194, Figure 10d,i,n), and 20 Hz (k = 0.259, Figure 10e,j,o). For reference, the values of the C_D, C_L, and C_M for a static model at varying fixed angles of attack with the same geometry [14] are superimposed in Figure 10a,f,k. The

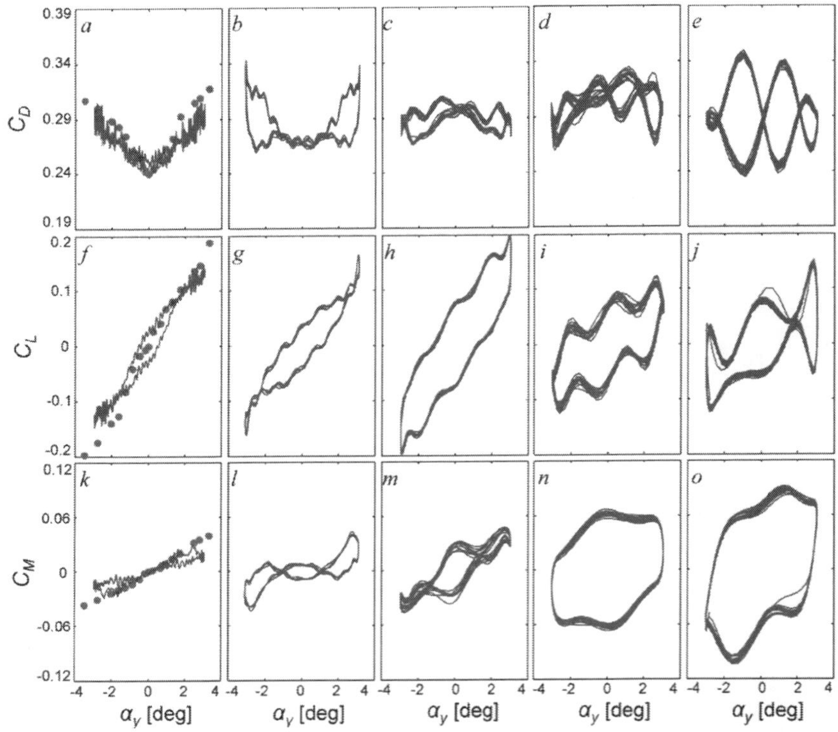

Figure 10. Drag (**a–e**), Lift (**f–j**), and Pitch (**k–o**) coefficients with pitching angle for the axisymmetric model with simple harmonic pitching at an amplitude of $\alpha_y = \pm 3°$ and oscillation frequencies of 1 Hz (**a**,**f**,**k**), 5 Hz (**b**,**g**,**l**), 10 Hz (**c**,**h**,**m**), 15 Hz (**d**,i,**n**), and 20 Hz (**e**,**j**,**o**) over a time interval of 1 s, using 100 phase averages, with a flow $Re_D = 2.3 \times 10^5$. The drag, lift, and pitch coefficients measured on the same model geometry at static angles by Lambert *et al.* [14] are shown in green symbols.

baseline C_D *vs.* αy presented in Figure 10a has a quadratic shape with a minimum of $C_D = 0.24$ at αy = 0 and a change in magnitude of around 25% (0.24 to 0.3). As frequency is increased, the quadratic functionality in the drag with angle becomes broken and the cycle average drag increases to 0.3 at 15 Hz (Figure 10b–d), and then reduces to 0.28 at 20 Hz (Figure 10e). It is noted that the shedding frequency of this model is expected to be around 100 Hz ($St_D = fD/U_0$ was measured to be ~0.2, for a similar axisymmetric body [16]), and the high frequency variations in C_D (as well as C_L and C_M) that are not present in the quasi-steady results are at frequencies of harmonics of the model motion near this shedding frequency. The baseline steady C_L *vs.* αy presented in Figure 10f is quasi-linear with a slight hysteresis and a lift slope of $C_L/\alpha y$ ~0.05/°. Upon increasing the pitching frequency to 10 Hz (Figure 10g,h), the hysteresis in the lift coefficient increases while the average lift slope remains similar, and at higher frequencies this slope diminishes to 0.03/° and 0.025/° at 15 and 20 Hz, respectively (see Figure 10i,j). The tendency of the C_L (and C_M) hysteresis to increase with frequency is attributed to unsteady effects of the flow (the magnitude of the wake's response time remains roughly similar as the model cycle time decreases). The quasi-steady C_M is shown in Figure 10k, with an increasing slope $C_M/\alpha y$ with frequency from 0.01/° at 1 Hz to 0.025/° at 20 Hz, with a large increase in the hysteresis of the response at higher frequencies (Figure 10l–o). It is noted that this "average slope" is defined as the slope of a linear fit through the dataset and does not take in account variations due to the hysteresis. In addition, having a positive average slope of C_M with αy implies that the model is unstable in pitch (the moment increases with an increase in the pitching angle), as expected.

The second motion investigated is dynamic $z = \pm 5$ mm ($z/D = \pm 0.056$) amplitude simple harmonic plunging, shown inFigure 11. Figure 11a–e show C_D, Figure 11f–j show C_L, and Figure 11k–o show C_M with the plunging coordinate, z, analogous to Figure 10, with the same flow conditions, frequencies, and timing. As the motion increases in frequency, the time averaged C_D stays around the same at 0.24, with the introduction of a higher frequency fluctuation (varying in magnitude from 0.01 to 0.08 in Figure 11a–e) attributed to the wake instability. The quasi-steady C_L is approximately invariant (Figure 11f), but has the introduction of a small negative slope at higher frequencies (C_L/z = 0.015/° at 10 Hz, 0.01/° at 15 Hz, and 0.015/° at 20 Hz) in addition to a large higher frequency fluctuation (also attributed to shedding). This negative average slope of C_L with z implies that the model is stable in plunge, as the aerodynamic force opposes the plunging direction. The quasi-steady C_M has slight variations, shown in Figure 11k, with negligible slope throughout its frequency, until the plunging frequency is 20 Hz, with the introduction of a C_M/z slope. This may be due to small deviations between the commanded zero pitching angle of the model and the realized pitching angle, as the pitching moment is very sensitive to pitching angle at this large unsteady frequency (compare Figure 10o and Figure 11o).

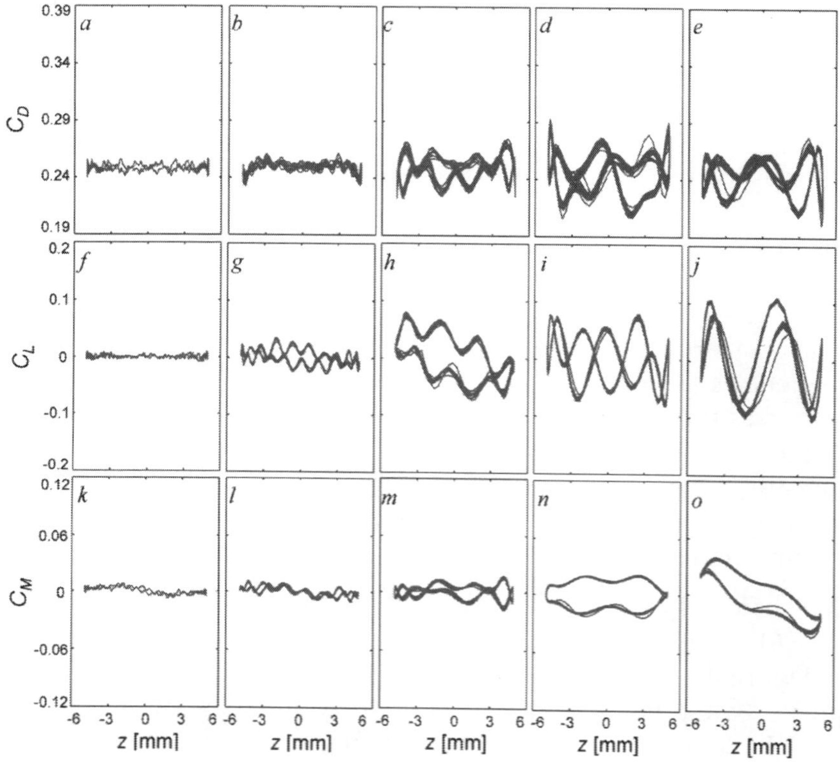

Figure 11. Drag (**a**–**e**), Lift (**f**–**j**), and Pitch (**k**–**o**) coefficients with simple harmonic plunge at frequencies of 1 Hz (**a**,**f**,**k**), 5 Hz (**b**,**g**,**l**), 10 Hz (**c**,**h**,**m**), 15 Hz (**d**,**i**,**n**), and 20 Hz (**e**,**j**,**o**) and an amplitude of $z = \pm 5$ mm, for the same conditions as in Figure 10.

The third and last canonical motion is dynamic $x = \pm 5$ mm ($x/L = \pm 0.030$) amplitude simple harmonic streamwise displacement, shown in Figure 12. C_D (Figure 12a–e), C_L (Figure 12f–j), and C_M (Figure 12k–o) are presented with the same frequencies, timing, and conditions as Figure 10 and Figure 11, for completion. For this motion, it is noted that both the motion and the freestream velocity are in the same direction, so the predominant

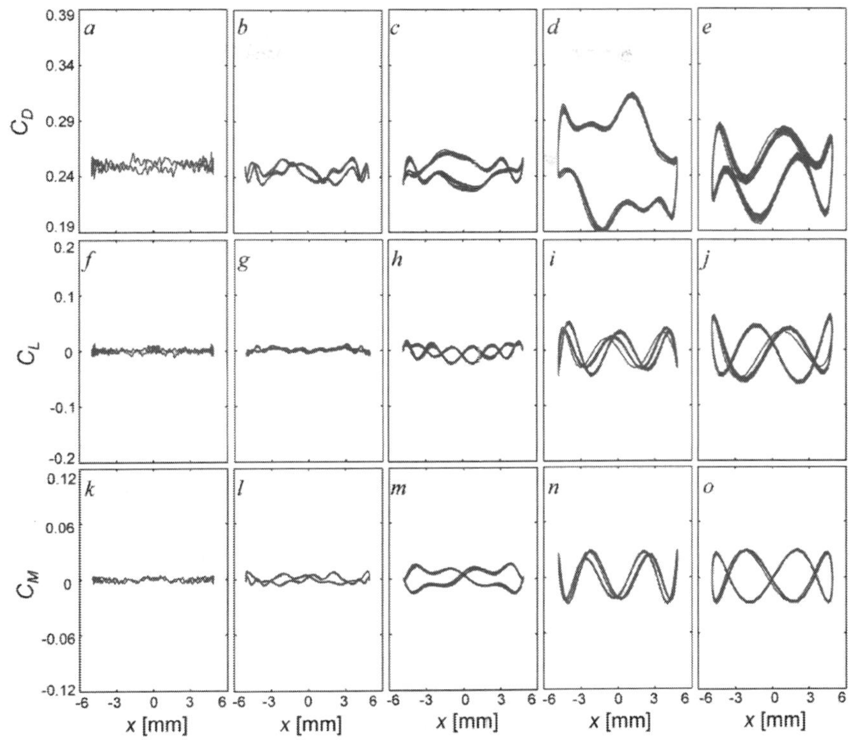

Figure 12. Drag (**a–e**), Lift (**f–j**), and Pitch (**k–o**) coefficients with simple harmonic streamwise displacement at frequencies of 1 Hz (**a**,**f**,**k**), 5 Hz (**b**,**g**,**l**), 10 Hz (**c**,**h**,**m**), 15 Hz (**d**,**i**,**n**), and 20 Hz (**e**,**j**,**o**) and an amplitude of $x = \pm 5$ mm, for the same conditions as in Figure 10.

changes occur in C_D, while the changes to C_L and C_M are attributed only to phase locking to the instability in the wake, and are approximately symmetric with zero time averaged values (Figure 12f–o). The time averaged value of the drag retains a value of approximately 0.24, with a growth in hysteresis at 10 Hz (Figure 12c), which maximizes at 15 Hz (Figure 12d), and then decreases at 20 Hz (Figure 12e). It is noted that for most of these drag measurements the average slope of the drag is zero, with the exception of 15 Hz where there is a slight slope of C_D/x of $-0.01/°$ which suggests this motion is neutrally stable (the aerodynamic drag is roughly invariant to changes in streamwise coordinate throughout most frequencies). This can be expected as this motion can be analyzed as a fluctuation of the freestream with a value of $2\pi fx = \pm 0.0314$ to 0.628 m/s, which should not induce large fluctuations in the steady drag coefficient, and the only induced effects are expected to be induced in the unsteady aerodynamics regime.

3.3. Wake Response

In order to understand the flow mechanisms that result in forces measured on the moving body mounted in this traverse, it is necessary to be able to measure or estimate the wake response around the moving model. The wake response can be measured relative to the real time states of the model as the traverse controller outputs an analog trigger to external devices at either fixed time or fixed trajectory offsets. This triggering system is used to trigger an external *LaVision* particle image velocimetry (PIV) system to capture instantaneous images of the wake at the aft end of the model in an interrogation region of $1.6D \times 1.2D$ ($3.2R \times 2.4R$) in a vertical region along the centerline of the model. Figure 13 shows six phase-averaged flow fields (each based on 300 individual realizations) for the $\alpha y = \pm 3°$ (Figure 13a,d,g,j,m,p), $z = \pm 5$ mm (Figure 13b,e,h,k,n,q), and $x = \pm 5$ mm (Figure 13c,f,i,l,o,r), with velocity field vectors overlaid on a raster plot of the cross-stream vorticity, ζ, at $Re_D = 2.3 \times 10^5$ ($U_0 = 40$ m/s). The frequency of these motions are chosen to represent unsteady wake dynamics at 20 Hz ($\tau = 0.05$ s), and the phases shown are equally spaced in time, t, throughout the cycles at: $t/\tau = 0$ (Figure 13a–c), 0.167 (Figure 13d–f), 0.333 (Figure 13g–i), 0.500 (Figure 13j–l), 0.667 (Figure 13m–o), and 0.833 (Figure 13p–r). For reference, the aft end of the model is superimposed on this data and shown in gray. The wake behind this model has three primary regions of interest: (i) the shear layer with the largest vorticity concentration; (ii) the inner wake with a reversed flow and smaller concentration of vorticity, which is bound by the wake stagnation point; and (iii) the boundary layer along the aft end of the body, which clearly indicates the time-resolved stagnation point on the body. For $\alpha y = \pm 3°$ pitch (Figure 13a,d,g,j,m,p), the chosen time scale leads to an unsteady flow response, where the wake development time scale is on the same order of the model motion. This leads to the wake response being out of phase with the model (note the wake is approximately centered between Figure 13a,d, and Figure 13j,m, respectively, implying an approximate phase lag of the wake response of about 30°) which is also commensurate with the hysteresis measured previously in C_L *vs.* αy and C_M *vs.* αy (Figure 10e,j,o). For $z = \pm 5$ mm plunge (Figure 13b,e,h,k,n,q), the wake response looks similar to pitching with a smaller deflection and a different phase lag (an easy way to observe this effect is to compare the development of the stagnation point on the model in Figure 13b,e,h,k,n,q toFigure 13a,d,g,j,m,p). For $x = \pm 5$ mm streamwise displacement (Figure 13c,f,i,l,o,r), the wake seems approximately unchanged and translates with the model with an addition of small fluctuations in the vorticity in the reversed flow region of the inner wake. In addition to Figure 13, videos are provided with the manuscript that depict the αy = ±3° pitch (Video S1), $z = \pm 5$ mm plunge (Video S2), and $x = \pm 5$ mm streamwise displacement (Video S3) wake dynamics at $f = 20$ Hz (seeSupplementary Materials). These results show that the present dynamic traverse is capable of effecting high frequency harmonic motions of the attached model, as well as resolving both the induced aerodynamic forces on a model and the corresponding wake response.

4. DISCUSSION AND CONCLUDING REMARKS

The present experimental work is focused on a design, manufacturing, and implementation of a novel 6-DOF traversing mechanism that enables realization of dynamic states of a model and characterization of its motion and force/moment responses. This traversing mechanism suspends a wind-tunnel model by eight wires with custom servo actuators. This mechanism is designed to move all the wires in patterns that can cause rotation in three independent axes: roll (8°), pitch (15°), and yaw (9°), as well as the respective independent translations: streamwise, cross-

Figure 13. Raster plots of the phase-averaged vorticity field with overlaid phase-averaged velocity vectors for simple harmonic motions (f = 20 Hz) in α_y = ±3° pitch (**a**,**d**,**g**,**j**,**m**,**p**), z = ±5 mm plunge (**b**,**e**,**h**,**k**,**n**,**q**), and x = ±5 mm streamwise displacement (**c**,**f**,**i**,**l**,**o**,**r**), at phases t/τ = 0 (**a**–**c**), 0.167 (**d**–**f**), 0.333 (**g**–**i**), 0.500 (**j**–**l**), 0.667 (**m**–**o**), and 0.833(**p**–**r**) with a flow $Re_D = 2.3 \times 10^5$.

stream, and vertical; all within a range of 50 mm at lower frequencies. The range of the traverse is dependent on the frequency of motions as well as the geometry of the model which is to be investigated. Each wire has an incorporated load cell that resolves the tension, and thereby captures the aerodynamic forces and moments on the model. Motion is executed by an in-house real-time controller that provides signals to the servo actuators as commanded. The executed motion is recorded in six degrees of freedom by an external *VICON* camera system (600 fps), which output is fed into the controller as a feedback signal to maximize the accuracy of motion. The real-time control system utilizes two *Quanser* data acquisition boards.

The traverse-driven model motion is tested in multiple degrees of freedom, and it could generate desired complex trajectories, comprised of the combined translational and rotational motions. These trajectories could be realized with minimal error compared to the commanded motion, being executed in a PID control loop in addition to having feedback from the motion analysis system. In the present work, a time trace of pitch and yaw commands out of phase are generated to demonstrate the robustness of this traverse controller in multiple degree of freedom motion at a representative "low" frequency of 1 Hz, and a representative "high" frequency of 10 Hz.

For the purposes of the traverse testing, an axisymmetric model is used as a case study. Three canonical motions of this model are chosen for investigating both the aerodynamic loads as well as the commensurate wake responses behind the model: sinusoidal pitch at ±3°, vertical plunge at ±5 mm, and horizontal displacement at ±5 mm. The corresponding measurements were taken at multiple motion cycle frequencies ranging from 1 to 20 Hz. Several known characteristics of this platform were used to validate the force measurements, such as its drag coefficient, its static values of drag, lift, and moment at fixed angles of attack, and its stability. The induced aerodynamic moment on the body undergoing pitching motion is of the same sense as the pitch angle, and is unstable. In contrast, the lift induced force on the body undergoing vertical plunge is of the opposite sense and is stable. It is also shown that the higher frequency motions excite force responses with frequencies that are of the order of the expected vortex shedding frequency of the model geometry. The ensuing wake responses are measured and analyzed, using a triggering system incorporated into the traverse trajectory controller and an external PIV system that resolved the wake development with a cycle frequency of 20 Hz. The test results imply that the developed support mechanism can be used for dynamic tests of various models, with time-resolved trajectory (and its derivatives), aerodynamic loads, and the flow dynamics.

ACKNOWLEDGMENTS

This work has been supported by Georgia Tech and the Army Research Office.

AUTHOR CONTRIBUTIONS

Thomas J. Lambert, Bojan Vukasinovic, and Ari Glezer contributed equally to this work, including conception, design, experiments, data analysis, interpretation, and writing.

CONFLICTS OF INTEREST

The authors declare no conflict of interest.

ABBREVIATIONS

The following abbreviations are used in this manuscript:

DOF degrees of freedom

PIV particle image velocimetry

CAD computer-aided design

NIR near infrared

UAV unmanned aerial vehicles

SLA stereo-lithography

AWG American wire gauge

EMI electromagnetic interference

DAQ data acquisition

UDP universal data protocol

PID proportional-integral-derivative

CV control volume

CG center of gravity

REFERENCES

1. Higuchi, H.; van Langen, P.; Sawada, H.; Tinney, C.E. Axial flow over a blunt circular cylinder with and without shear layer reattachment. *J. Fluids Struct.* **2006**, *22*, 949–959.
2. Higuchi, H.; Sawada, H.; Kato, H. Sting-free measurements on a magnetically supported right circular cylinder aligned with the free stream. *J. Fluid Mech.* **2008**, *596*, 49–72.
3. Achenbach, E. Experiments on the flow past spheres at very high Reynolds numbers. *J. Fluid Mech.* **1972**, *54*, 565–575.
4. Taylor, G.S.; Gursul, I.; Greenwell, D.I. Investigation of support interference in high-angle-of-attack testing. *J. Aircr.***2003**, *40*, 143–152.
5. Taylor, G.S.; Gursul, I. Support Interference for a Maneuvering Delta Wing. *J. Aircr.* **2005**, *42*, 1504–1515.

6. Griffin, S.A.; Crooks, R.S.; Mole, P.J. Vane Support System (VSS) A New Generation Wind Tunnel Model Support System. In Proceedings of the 29th Aerospace Sciences Meeting, Reno, NV, USA, 7–10 January 1991.

7. Pattinson, J.; Lowenberg, M.H.; Goman, M.G. Multi-degree-of-freedom wind-tunnel maneuver rig for dynamic simulation and aerodynamic model identification. *J. Aircr.* **2012**, *50*, 551–566.

8. Tuttle, M.H.; Gloss, B.B. *Support Interference of Wind Tunnel Models: A Selective Annotated Bibliography*; NASA-TM-81909-SUPPL; NASA Langley Research Center: Hampton, VA, USA, 1 March 1981.

9. Bacon, D.L. *Model Supports and Their Effects on the Results of Wind Tunnel Tests*; NACA Technical Notes 130; National Advisory Committee for Aeronautics: Hampton, VA, USA, 1 February 1923.

10. Ericsson, L.E.; Reding, J.P. Review of support interference in dynamic tests. *AIAA J.* **1983**, *21*, 1652–1666.

11. Beyers, M.E. Unsteady wind-tunnel interference in aircraft dynamic experiments. *J. Aircr.* **1992**, *29*, 1122–1129.

12. Abramson, P.; Vukasinovic, B.; Glezer, A. Direct measurements of controlled aerodynamic forces on a wire-suspended axisymmetric body. *Exp. Fluids* **2011**, *50*, 1711–1725.

13. Abramson, P.; Vukasinovic, B.; Glezer, A. Fluidic control of aerodynamic forces on a bluff body of revolution. *AIAA J.***2012**, *50*, 832–843.

14. Lambert, T.J.; Vukasinovic, B.; Glezer, A. Active decoupling of the axisymmetric body wake response to a pitching motion. *J. Fluids Struct.* **2015**, *59*, 129–145.

15. Hoerner, S.F. *Fluid Dynamic Drag*, 3rd ed.; Hoerner Fluid Dynamics: Bricktown, NJ, USA, 1965.

16. Rigas, G.; Oxlade, A.R.; Morgans, A.S.; Morrison, J.F. Low-dimensional dynamics of a turbulent axisymmetric wake.*JFM Rapids* **2014**, *755*, R5.

5

CHAPTER

Design and Development of a Gas Dynamics Facility and a Supersonic Wind Tunnel

N. A. Ahmed[1]

[1] Aerospace Engineering, School of Mechanical and Manufacturing Engineering, University of New South Wales, Sydney, NSW, Australia

1. INTRODUCTION

The design of a Supersonic Wind Tunnel is complex, expensive and time consuming. One of the pre-requisites of such a facility is the availability of compressed air necessary to generate the required speed.

In this Chapter, the design and construction of the basic gas dynamics facility (Fig. 1) is described first in Part I followed by that of a blow down type supersonic wind tunnel (Fig. 2) in Part II. The two facilities are currently in operation at the School of Mechanical and Manufacturing Engineering of the University of New South Wale.

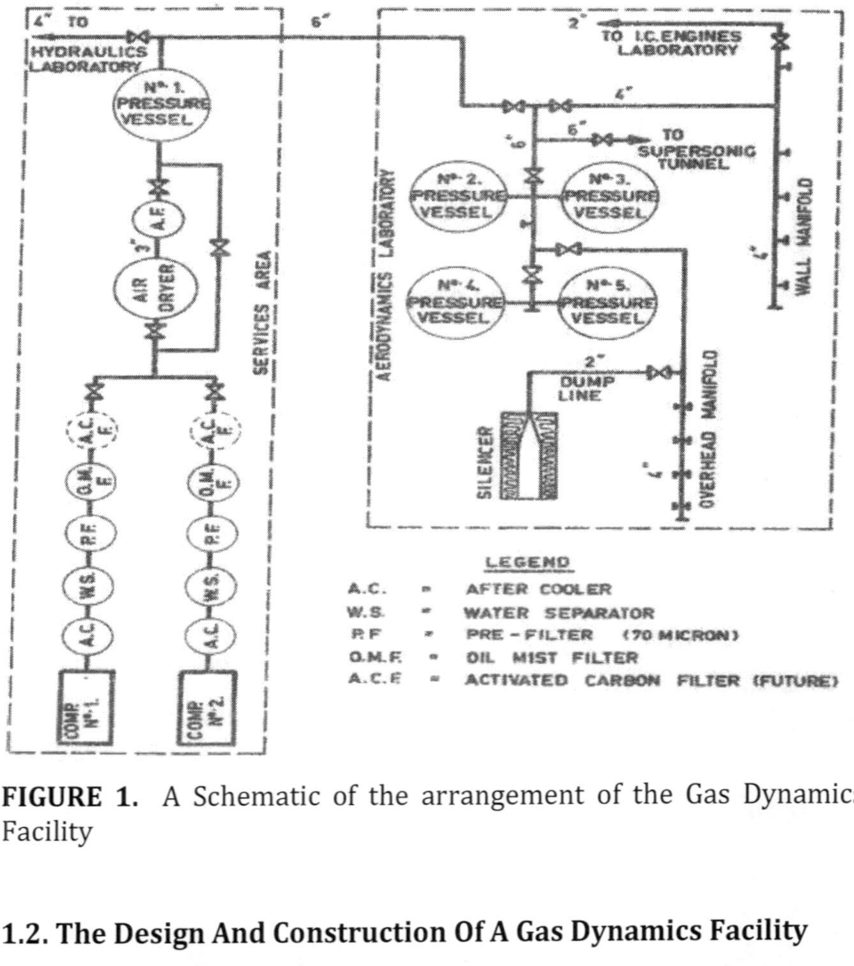

FIGURE 1. A Schematic of the arrangement of the Gas Dynamics Facility

1.2. The Design And Construction Of A Gas Dynamics Facility

The Gas Dynamics facility consists of a large capacity compressed air plant that involved the installation in the Aerodynamics Laboratory of the University of New South Wales of a compressed air plant on the floor and the construction of an overhead structure of four identical 200 cubic feet capacity storage pressure vessels.

The design was initiated by setting a requirement of continuous mass flow rate of about 1 lb/sec. For continuous operation, the system pressure was set to about 100 psig. The gas dynamics facility was also required to provide air supply to a 3.5 inch x 4 inch supersonic wind tunnel, capable of operation of up to Mach 3.5.

FIGURE 2. A side view of the 5 ½ inch x 4 inch Supersonic Wind Tunnel

A brief description of the gas dynamic facility is given next.

1.2.1. Compressed air plant

Compressors

From a consideration of vibration and intake resonance of the machines and also to provide significantly larger outputs per unit cost, it was decided to use rotary compressors. Two Holman RO600S screw type compressors, each rated at 600 cubic feet per minute of free air each and capable of operation separately to provide a mass flow of 0.75 lb/sec or in parallel for an output of 1.5 lb/sec were acquired. Each compressor maximum pressure ratings is 100 psig for continuous operation and 115 psig for intermittent operation such as that required for use with supersonic wind tunnels. Each unit is driven by a 150 HP 1440 RPM induction motor controlled by an auto-transformer started capable of up to 15 starts per hour. Each compressor unit was installed on 'Vulcascot' anti-vibration matting and was isolated from the discharge pipe work by means of a flexible pipe work connector. As an additional precaution, the first length of outlet pipe work to the after coolers was supported on anti-vibration matting. The result is that with both compressor operating, virtually no vibration is transmitted to the Laboratory building. A schematic of the compressed air plant is shown in Fig 1.

Control of the compressor output pressure is by either an automatic stop-start system or a constant speed uploading system operating between pre-set pressure limits. In operation, the constant speed uploading mode has been most frequently used but the original pneumatic system supplied with the compressors for this purpose proved to be unreliable. Subsequently, this was replaced with an electrical system utilising an electric control pressure gauges. This system has proved to be very satisfactory in operation and enables repeatable and readily varied settings of cut-in and cut-out pressure to be obtained with differentials as small as 2 psi.

The compressors are cooled by oil injection and lubricated by the same oil pressurised from a small pump. The cooling/lubricating oil is stored in a 40 gallon tank and cooled by an oil/water heat exchanger. The air, after compression, passes through a multi-stage reverse flow oil separator with absorbent filters so as to remove most of the oil present. Claimed oil consumption is one gallon per compressor per 400 hours of operation. Fig. 3 shows the Compressor of the gas dynamics facility.

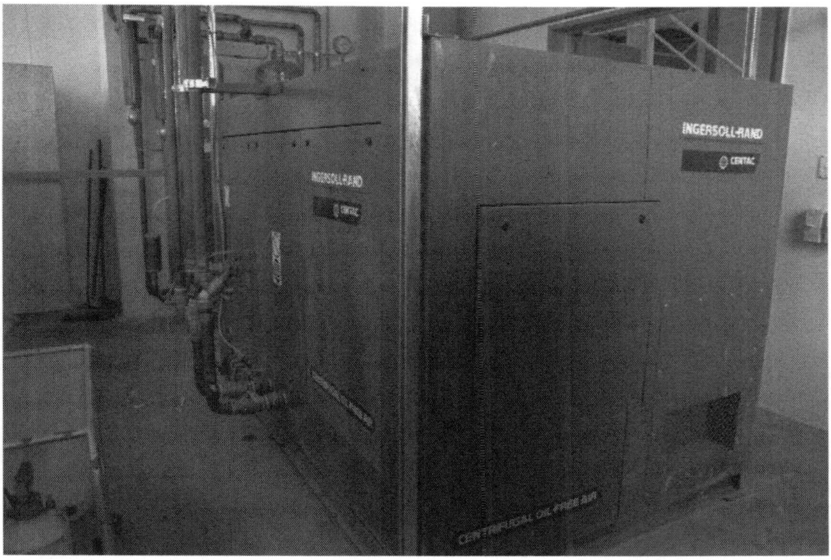

FIGURE 3. The Compressor of the Gas Dynamics Facility

1.2.2. Filtration system

After compression, air from each compressor is passed through individual shell and tube after coolers, centrifugal action water separators, pre-filters and oil-mist filters. The pre-filters consist of a centrifugal action separator combined with a 70 micron porosity, sintered bronze filter element. The oil mist filters are a proprietary design of Daltech Engineering Incorporated, USA, and consist of a stainless wire wool pre-filter and a chemical absorption type main element. These filter units are claimed to have a 99.4% filtration efficiency for all oil particulates down to 0.5 microns. Both pre-filters and oil-mist filters operate with a 'colour change' principle in that, as the element becomes saturated with oil, its colour changes from light pink to bright red. All filters, after coolers and water separators are fitted with automatic water drains.

1.2.3. Air Drier

The air drier (Fig. 4) is a proprietary designed manufactured in Australia by B.C. Johnson Ltd. It is designed for inlet air conditions of

1.5 ib/sec at a pressure of 115 psig, a maximum temperature of 105^{0} F and an ambient relative humidity at compressor inlet of 100%. The dryer is required to provide an outlet humidity equivalent to -50 ^{0} F at atmospheric pressure after a two hour drying cycle.

FIGURE 4. Air Drier System

The air dryer is of the single tower at a larger stage. Regeneration is accomplished by air pre-heated in an 80 kW electrical heater and forced through the dryer stack in a counter-flow direction by a centrifugal blower. The desiccant employed is 800 lb of activated alumina in a 2 ft diameter by 4.5 ft high bed, preceded by 80lb of buffer desiccant whose purpose is to prevent damage to the main desiccant by liquid carry over from filtration equipment. The dryer is fitted with a water coil for cooling the desiccant bed after regeneration and a felt pad and fibreglass after-filter is installed to prevent any carryover of desiccant dust into the storage vessels.

In operation, dew points of as low as -80 ^{0} F have been obtained after regeneration. With one compressor in operation, drying times of up to six hours have been achieved, although a higher final dew points than -50 ^{0} F regeneration time is about four hours for heating and four hours for cooling.

1.2.4. Air storage vessels

Four storage vessels (Fig.5), each 5 ft diameter by 11 ft 6 ins overall length and designed for a working pressure of 130 psig have been installed with a total storage capacity o f800 cubic feet as mentioned earlier were placed in an overhead structural steel support in the Aerodynamics laboratory near which a supersonic wind tunnel was built.

FIGURE 5. Air Storage Vessels

1.2.5. Air conditions 105 ^{0} f at atmospheric pressure

Air conditions in the storage vessels are a volume of 800 cubic feet at ambient temperature and dried to a dew point the equivalent of at least -50 ^{0} F at atmospheric pressure. For intermittent operation, maximum pressure is 125 psia. Maximum intermittent mass flow rates are in the region of 10 to 20 lb/sec to limit the system pressure loss to approximately 5 psi. Maximum continuous mass flow is 1.5 lb/sec with a pressure not exceeding 110 psia.

Because the pressure available for supersonic tunnel injection is comparatively low, care was taken in the design and piping and filtration equipment to reduce pressure losses. The overall pressure drop between compressor outlet and storage vessels has been kept between 3 and 5 psi depending upon filter condition.

1.2.6. Air distribution manifolds

The supply pipe work inter-connecting the five pressure vessels is of 6 inch inside pipe diameter pipe. There is a 6 inch branch to the supersonic tunnel. Gas dynamics rigs in the Aerodynamics Laboratory are supplied from two 4 inch pipe manifolds, one wall mounted and the other suspended from the ceiling. A four inch line, reducing to 3 inch, supply air to the Hydraulics Laboratory.

Maximum intermittent flow rates are about 10 lb/sec through the 6 inch branch supplying the supersonic tunnel and 4 lb/sec through the 4 inch manifolds. At these flow rates, the pressure drop between reservoirs and manifold outlets does not exceed about 3 psi. The hydraulics Laboratory supply system permits an intermittent flow rate of about 5 psi.

A 2 inch dump line is provided, together with a control valve and attenuating duct silencer to empty the storage vessel contents or to permit an adjustable air bled for stabilization purposes.

2. THE DESIGN AND CONSTRUCTION OF A BLOW DOWN TYPE SUPERSONIC WIND TUNNEL

2.1. Design Of Supersonic Tunnel Components

Some of the details of the design of the various tunnel components are described in the following sections. The aerodynamic configuration finally selected for the tunnel is shown in Fig. 6.

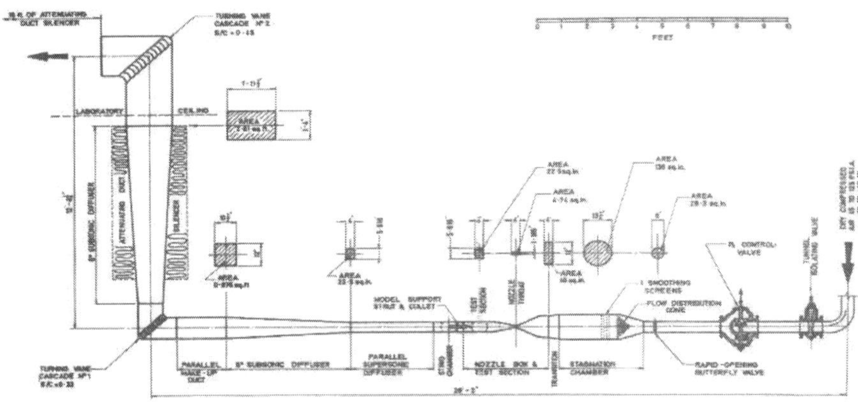

FIGURE 6. A Schematic of the 5 ½ inch x 4 inch Supersonic Wind Tunnel

2.1.1. Intake piping and control valves

The following requirements received attention:

- Reduction of pressure loss
- The need to supply a uniform airflow free of pulsation to the stagnation chamber

In addition to the selection of a 6 inch diameter for the tunnel intake pipe work, as mentioned in section 1.6, extra measures to reduce pressure losses and ensure flow uniformity were fitting of splitter vanes to all piping bends and tees in the final run to the tunnel. The design data of Ito (ref 29) was for this purpose.

Three valves were fitted for the flow control (Fig. 6). The first, a 6 inch gate valve, serves merely as the tunnel isolation valve and a backup shut-off valve. The second valve, downstream of the gate valve, is the stagnation pressure control valve. This is followed by the quick opening valve which is located at the inlet to the tunnel stagnation chamber.

The stagnation pressure control valve is a 6 inch double seat Fisher Governor Company valve with pneumatic cylinder actuation. The valve is of the 'Vee-pup' type which has equal percentage flow characteristics. This characteristic restricts the rate of valve spindle movement which would otherwise be necessary when the pressure drop across the valve decreases towards the end of a tunnel run. Control of the valve opening is by means of a standard 3 to 15 psi regulator located at the tunnel control panel. This regulator, acting on the positioned, supplies air at up to 100 psi to the piston of the cylinder actuator. The 6 inch valve size was selected to limit the wide open pressure drop to less than 3 psi. Preliminary design estimates indicated that the pressure drop in a 4 inch diameter control valve would have been in the region of 15 to 20 psi. The double seat valve configuration ensures reasonable symmetry in the airstream approaching the stagnation chamber and assists in reducing pressure pulsations.

The quick-opening valve is a 6 inch diameter Fisher continental rubber seat butterfly valve with pneumatic cylinder actuation and a stroking time of less than one second. It is the last component in the 6 inch line before the stagnation chamber. The position gives the most rapid possible tunnel start using standard valves. Pressure loss is about 0.1 psi. An important advantage in operation of the tunnel is gained by locating the quick-opening valve downstream of the stagnation pressure control valve as the latter can then be correctly pre-set to the required starting pressure. The valve disc position when wide open ensures flow symmetry to the stagnation chamber. The quick opening valve is actuated by a solenoid operated air valves which are, in their turn, controlled by a push button solenoid circuit on the control panel. An electrical interlock is provided so that the tunnel cannot be started until the test section access facility has been securely closed. Tunnel operation may be stopped either at the control panel or from a wandering lead and control box operated by the tunnel engineer.

Maximum Mach number in the intake pipework, excluding 'jetting' from the stagnation control valve, is of the order of 0.1. The maximum calculated pressure loss from the pressure vessels to the stagnation chamber inlet is approximately 5 psi.

The Stagnation pressure control system was deliberately chosen to be manual in orer to simplify control although a hybrid system could be incorporated at a later stage if desired. This system would consist of manual start and initial stabilisation with switch over to automatic operation once the stagnation pressure has stabilised. Some of the problems of supersonic tunnel automatic stagnation pressure control have been discussed by Pugh and Ward [1] and Conolan [2].

2.1.2. Stagnation chamber

The stagnation chamber has an inside diameter of 13.5 inches and is connected to the 6 inch inlet pipe work by a 300 conical rapid expansion. Flow stabilisation and smoothing devices consist of a conical perforated plate and flow smoothing and turbulence reduction screens (Fig. 6). A parallel settling length is provided downstream of the screens and is fitted with stagnation temperature and pressure tappings.

The mean velocity at the screens is approximately 20 ft/sec for Mach 3 operation. The chamber is also sized to permit operation down to Mach 1.5 using the same test section area when the velocity at the screens could increase to about 70 ft/sec. This remains within the range of 10 to 80 ft/sec as recommended by Pope [3].

The perforated cone has an apex angle of 900 and is manufactured from a ¼ inch plate with 3/8 inch diameter holes on 9/6 inch centres. The perforations have an open area ratio of 40%. The mean Mach number through the perforations under the worst conditions, which occur at the lowest test section Mach number, is less than 0.1. The cone is welded into the wide angle expansion. In operation, it appears to have eliminated any pressure fluctuations generated by the stagnation pressure control valve as well as assisted in 'filling' the wide angle diffuser.

The four stainless steel flow smoothing screens are of 24 mesh by 34 gauge wire and have an open area ratio of 49.5%. The screens are fixed in individual aluminium retaining ring frames. These frames are clamped together by long bolts passing through large frames attached to the rapid expansion and settling length sections of the stagnation chamber. The individual frames are spigoted together to ensure internal surface alignment and are sealed by O rings at each joint.

The parallel settling length downstream of the screen is 1500 screen wire diameters long, or a length of approximately 18 inches.

A two dimensional contraction and section change transition region is provided at the downstream end of the settling chamber. This region has a rectangular outlet area of 12 inch x 4 inch and a circular 13.25 inch diameter inlet section. The area contraction ratio is 2.9:1. A further two-dimensional contraction, of ratio 10:1, is built into the nozzle blocks to contract the airstream to a sonic throat 1.185 inch high by 4 inches wide. The method of contraction design presented by Gibbings [4] is recommended. This method is also applicable to contractions in which there is an appreciable axial trigger between the plan and elevation profiles.

In operation, the stagnation chamber provides a steady pressure with accuracy of control of 1% or better. Estimated pressure loss for the stagnation chamber flow smoothing devices was 2 psi approximately, making a total estimated loss between pressure vessels and stagnation pressure measurement station of nearly 7 psi. In operation, this pressure loss varies between 7 -10 psi.

2.1.3. Nozzle box and test section

The nozzle box is of conventional construction and is manufactured from steel plate with internal surfaces ground after welding and stress relieving. Heavy stiffening ribs prevent deformation under pressure forces, particularly in the throat region. Dowels are fitted to ensure accurate and repeatable alignment of adjacent parts. Circumferential 'O' ring seals are provided at each end of the nozzle box. One side wall opens downwards on hinges to facilitate nozzle block changes. The complete assembly of nozzle box and settling length section of the stagnation chamber can be moved on rollers to permit easy screen removal. The rollers are brought into operation by four jack screws. An axial movement of 3 inches in the downstream direction is possible.

There are circular Schlieren window positions in the nozzle box walls, one pair at the throat and one pair at the test section. The windows have a clear diameter of 7.5 inches and thickness of 1 inch and are held in a sub-frame which is fixed to the tunnel by a clamping ring. This arrangement permits easy removal and rotation of each window. Rotation of the window assembly permits selection of the optimum orientation for optical characteristics of the glass fitted. This allows the use of cheap and comparatively low grade plate glass. The glass is sealed to the sub-frame with Dow Corning Silastic 732 RTV silicon rubber compound. A special jig has been developed for window assembly which ensures that the glass is completely 'floating' in Silastic and is also flush to within 0.001 inches with the frame edges. A set of high quality glass windows obtained from Optical Works, UK are also available for specially sensitive Schlieren applications.

The supersonic nozzle blocks are manufactured from extruded AA28S aluminium alloy. The contours were generated by a programmed 'Hydroptic jig borer and were finally hand finished to remove machining marks. Each block is fitted with a continuous, circular cross section rubber seal which runs as close to the contoured surface as is possible using straight line approximations. The nozzle blocks are held in the nozzle box with the bolts passing through the top and bottom walls of the box into barrel nuts inside each block. Location is by integral machined pads on the basis of each block. It is now realised that the provision of separate ground pads on the base of each nozzle block configuration would have permitted the fitting of permanent shims clamped between pad and block, thereby simplifying the fitting and accurate setting up of each nozzle block configuration within the nozzle box.

The nozzle block co-ordinates are those derived by McCabe [5] for operation at Mach 3 in a nominal 5.5 inch x 5.5 inch test section. The design method used by McCabe divides the nozzle inviscid core flow into five main regions, as follows:

- A subsonic contraction
- The high subsonic, low supersonic throat region
- An initial expansion region wherein the contour slope increases to its maximum value

- The straightening or 'Busemann Region'
- The parallel flow or test section region

McCabe also divides each of the 3rd and 4th regions into further regions so as to take advantage of more precise computational methods and eliminate discontinuities in nozzle curvature. The nozzle contour boundary layer corrections are based on the data of Sibulkin [6] for the throat region, with corrections on the contoured and flat side walls obtained from the data of Rogers and Davis [6]. Because the nozzle contours so derived did not have zero slope at the test section location, a cubic curve was fitted to permit smooth transition to the parallel block region downstream of the test rhombus. The boundary conditions were continuity of ordinate, slope and second derivative at the upstream and zero slope at the downstream end. Before feeding to the jig borer, the complete set of nozzle block co-ordinates was smoothed by computer using a 6th order polynomial. Data on nozzle profile design may be obtained from Refs 8-15.

2.1.4. Diffusers and model support system

Downstream of the nozzle box are a string chamber, supersonic diffuser, first stage subsonic diffuser, parallel make up duct and a set of cascade turning vanes. The complete assembly downstream of the nozzle box to the corner cascade is mounted on a flanged wheel and rail system and may be moved 18 inches in the downstream direction away from the test section. This allows easy access for model changes. The face of the sting chamber is aligned with, and closed against, the nozzle box pressure seal by four dowel pins and tapered locking wedges. Four cam clamps are used to close the pressure seal between the cascade corner and the vertical second stage subsonic diffuser.

The sting chamber can be removed entirely, if required. It is fitted with a pair of circular openings of the same diameter as those in the nozzle box side walls and, can, therefore, accept the interchangeable Schlieren windows or metal window blanks. A side mounted sting may be fitted to either of the side window blanks. A vertical, full span, wedge nose strut is also in use as a model sting support.

The optimum design of a supersonic diffuser for a small wind tunnel presents a difficult problem as it must operate over a wide range of inlet flow conditions and Reynolds numbers. Moreover, at Mach numbers in excess of 2, optimum starting and running diffuser geometries diverge significantly. It is assumed in the following discussion that the supersonic diffuser is followed by a subsonic diffuser of small divergence angle.

In general, the airstream entering the diffuser is highly turbulent because it contains the wake flow from the test section model and its support system. Moreover, diffuser performance is influenced by boundary layer effects which are a function of tunnel Reynolds number.

Supersonic diffusers for small wind tunnels may be one, or a combination of the following types:

- Fixed contraction with constant area second throat
- Variable contraction with variable area second throat
- Constant area duct
- Oblique shock diffuser with centre body.

The diffuser throat must be sized must be to 'swallow' a normal shock when starting at the design Mach number in a second throat type of diffuser. Faro [11] obtained the theoretical minimum starting contraction ration required, together with experimental points from several sources. For Mach numbers in excess of 2, the experimental values are lower than the theoretical values. It is suspected that this is due the starting shock passing through the second throat before the design Mach number has been reached. The experimental data of Lukasiewicz [16] indicates that at the area ratio of A_2/A_1=0.7, which is required for stating at Mach 3, the starting pressure ratio requirement of a fixed geometry second throat diffuser, might be 0 to 30% less than that required by an optimum length constant area diffuser. If a variable geometry second throat diffuser is used, significant reductions in optimum running contraction ratio may be obtained particularly at Mach numbers in excess of 2.

Variable geometry diffusers, set at optimum running contraction ratios after starting, enable reductions in running pressure ratio of up to about 45% when compared with optimum length constant area ducts [17]. This reduction is achieved, however, at a considerable increase in complexity of construction and operation when compared with the simple constant area diffuser.

The advantages of the fixed and variable area second throat diffusers over the constant area diffuser are less certain when allowance is made for the presence of the model and its support system within the test section. Faro [11] states that in nearly all cases the presence of a model reduces supersonic diffuser efficiency. This is confirmed by De Leo and Huerta [17] who found that a 5% blockage model increased starting and running pressure ratios by approximately 7 and 13 % respectively for optimum empty tunnel diffuser geometry and Mach numbers of 2.5 and 3.4. However, experiments have also shown that the presence of a model tends to reduce tunnel instability through stabilisation of the diffuser shock system. This appears to be due to interaction between the wall reflections of the model bow wave and the diffuser shock system. In summary, therefore, it seems that the second throat diffuser types do not offer significant advantages over the constant area duct for small supersonic tunnels. This conclusion does not apply to very large intermittent type tunnels, however, as then a reduction in running pressure ratio could lead to significant increases in tunnel run time for a given pressure storage capacity or reduction in capital cost of the storage system for a given run time.

The supersonic diffuser of the University of New South Wales is a constant area duct system although the sting strut, when used probably acts as a centre body with oblique shock diffusion. Although such a constant diffuser is a dissipative system, it does have the considerable advantage of being stable over a wide range of inlet flow conditions and of being simple to construct and maintain.

The only design decision required for a constant area supersonic diffuser is the optimum length of duct. In such a diffuser the normal shock appears as a shock system which is strongly affected by the state of the boundary layer. Experimental dat quoted by Lukasiewcz [16] confirms that for best efficiency the shock compression process should be completed in the constant area duct and not in the divergent subsonic diffuser downstream of the supersonic diffuser. Faro [11] illustrates the gain in isentropic efficiency with increasing length to height ratio for a constant area duct at Mach 2. The significant reduction in operating pressure ratio with increasing length of parallel duct may be seen in curves plotted for length to height ratios of 0, 2 and 7. Further design data for constant area diffusers is demonstrated by Faro [11]. Two points are noted in connection with his work:

- The Mach number M_{av} is the average Mach number at the supersonic diffuser inlet and would be less than the test section Mach number because of the presence of the model and its support system and boundary layer growth between test section and diffuser.
- The effect of the free stream Reynolds number is not accounted for. Some indication of the reduction in the length of the shock compression system at high Reynolds numbers may be obtained from [16]. This data is applicable to a Mach number of approximately 2.

Faro [26] indicates that a single wedge such as the leading edge of a sting support strut may be used to provide an oblique shock system which will improve diffuser efficiency over the simple normal shock case. The benefits for this type of device are limited, however, is that high efficiency can only be obtained with a large number of oblique shocks which in turn implies design for a specific Mach number and thus a narrow range of off-design conditions. The simple strut type oblique shock generator gives moderate efficiency gains over a wider range of Mach numbers.

To summarise, little data for the design of constant area supersonic diffusers or for the effect of a model and strut system on diffuser efficiency can be found. The available information suggests that:

- The shock system compression process should be completed within the parallel diffuser duct for best efficiency
- The optimum length of parallel duct required to complete the compression process is a strong function of Mach number and Reynolds number. This length is probably within the 5 to 12 diffuser heights of Mach numbers of 1.5 to 3.5 and Reynolds numbers of 2×10^5 to 6×10^6. Design data for supersonic diffusers may be obtained from Refs 16-22].

The supersonic diffuser of the University of New South Wales tunnel is a parallel wall rectangular duct fabricated from 4 inch x 1 inch extruded aluminium bar top and bottom walls and 0.5 inch aluminium plate side walls. The top and bottom walls may be easily replaced with a set of contoured blocks so as to provide a fixed area second throat, if so desired. The parallel diffuser length is 8.4 diffuser heights from the rear of the model support strut and 11.4 heights from the end of the supersonic nozzle with model support system removed. A removable parallel subsonic make-up duct permits the fitting of an additional 4.3 diffuser heights. The window blanks alongside the model support strut also allow for experiments involving 'de-blocking' of the area around the strut.

The design of the subsonic diffuser was straightforward. The data of Lukasiewicz [16] indicates that, in the Mach number range 0.4 to 0.9, total pressure recovery is virtually constant at about 0.88 for open ducts without models and that the diffuser divergence angle should be less than 7^0. Data on subsonic diffuser design is available from Refs 14 and 16.

There are two stages of subsonic diffusion separated by the corner cascade (Fig. 6). The first stage of subsonic diffusion is separated by the corner cascade (Fig. 6). The first stage has an area ratio of 5.6 and divergence angle of 60. Maximum Mach number at the subsonic diffuser exit is approximately 0.13. The diffuser is constructed from 3/16 inch steel plate reinforced at 6 inch x 3 inch centres with 1 x 0.25 inch flat bars on edge.

The second stage subsonic diffuser has a 60 divergence angle and 3:1 area ratio. Maximum Mach number at exist to this diffuser is approximately 0.04. The diffuser is manufactured from plywood and incorporates part of the tunnel silencing system. The corner cascade utilises sheet metal circular arc turning vanes.

2.1.5. Silencer

Preliminary investigations on an existing M3.5, 4 inch diameter conical nozzle indicated that the noise level for an unsilenced tunnel would be unacceptably high at about 120 db in the frequency band of 100 to 2000 Hz. Accordingly, a silencer was designed for the supersonic tunnel to the following requirements:

- Noise reduction to about 80 dB in the 100 to 2000 Hz band
- Low pressure loss
- Ease of construction and low cost

After investigation, an attenuating duct design was chosen as best fulfilling these requirements. This type of silencer requires an absorbent material as dense as possible with a thickness of 2 inch to 12 inches to absorb the low frequency noise below 500 Hz. Attenuation at the lower frequencies is increased considerably by the use of a perforated duct facing material having about 3 to 10% open area perforations. Low frequency attenuation is further assisted by providing airspace behind the absorbent material and increasing the amount of absorbent around the duct periphery. When compared with splitter type duct attenuators, low frequency attenuation can be improved by arranging a given amount of attenuation material such that it forms thick layers. This latter arrangement gives a lower peak but better average attenuation over the 100 to 1000 Hz frequency band. Design information can be found in the literature [23-29].

The silencer for the University of New South Wales tunnel is constructed in two sections: the first of which is built around the second stage subsonic diffuser. The first section comprises 6 to 12 inch thickness of polythene wrapped rockwool batts and loose rockwool fill around around all four sides of the diffuser. The rockwool density varies from 4 to 6 lb/cubic feet for the loose fill. The duct interline is surfaced with 3/16 inch thick perforated plywood and the outside of the silencer is sealed with 1 inch thick, exterior quality waterproof plywood. Both internal and external surfacing materials are heavily glued, screwed and nailed to substantial connecting framing. The second section of the silencer, which is 16 ft long is a rectangular duct lined on two sides with 6 inch thickness of rockwool batts backed by a 3 inch airspace. The remaining two sides of this duct are 1 inch thick exterior plywood. Other constructional details are similar to those of the first section silencer. The second diffuser section is run in the laboratory ceiling space and is supported from the roof structure on 'Silentbloc' vibration isolators.

Initial tests on completion of the tunnel indicated a large direct sound transmission through the walls of the first stage subsonic diffuser. This was found to be caused by high frequency resonance of the 3/16 inch thick flat steel plate walls. The vibration was almost completely eliminated and the noise level reduced by decreasing the spacing of the existing 1 inch x 0.25 inch stiffening bars from approximately 12 inch x 6 inch to 6 inch x 3 inch centres as described in section 4.4.

In the final form, the silencer has reduced the noise level in the vicinity of the tunnel to about 75 to 90 dB, for the 100 to 2000 Hz band, depending to some extent upon the operating stagnation pressure. It is estimated that the duct silencer provides an attenuation of about 2 to 3 dB per foot of length in the frequency range 100 to 1000 Hz.

2.1.6. Instrumentation

The tunnel stop-start system has been briefly described in section 2.1.4.

Tunnel stagnation pressure is read on 0.15% accuracy, temperature compensated, absolute pressure 'Heise' test gauge and recorded by a pressure transducer having 0.1% combined non-linearity and hysteresis. The transducer output can be displayed directly in psia on an 11 inch 'Honewell' strip chart recorder. The control panel is provided with an electrically actuated pneumatic calibration circuit which connects the stagnation pressure transducer and test gauge in a closed system. This circuit has an electrical override if the tunnel is started.

Stagnation temperature instrumentation consists of an exposed-junction 'BLH' micro-miniature thermocouple connected to an 11 inch strip chart recorder and reading directly in 0F. Bothe stagnation temperature and pressure recorders contain electrically operated chart speed-up facilities which automatically increase the chart speed by a factor of 60:1 when the tunnel run is started. A typical speed change is from 10 inches per hour to 10 inches per minute. Both chart recorders are provided with event markers which are connected into the tunnel timing circuit. The circuit operates an electrically actuated second timer which is controlled from a timer switch in the remote control box on a wandering lead. The box also contains the tunnel stop switch and a pressure clamp switch. The event markers are automatically actuated at the start and stop of a timing run. The wandering lead control box enables one man to control the run and monitor Schlieren and instrument read out.

Data acquisition is by conventional multi-manometers, pressure gauges or a range of flush diaphragm transducers. For calibration, the data transducers can be connected into a closed system with a 'heise' stagnation pressure gauge. This only requires operation of a control panel push-button. Read out equipment for the transducers are EAI and Solartron multi-channel data loggers and a 6 channel pen recorder.

Tunnel air flow calibration equipment has been designed and built in accordance with the data published by Anderson [16]. Air humidity is measured with a Casella-Alnor dew point meter which can measure dew points to -80^{0}F with an accuracy of 3^{0}F.

A two-mirror, parallel path, 96 inch focal length, 7.5 inch aperture Schlieren system is currently in use with the tunnel and associated gas dynamics rigs. A second 60 inch focal length system is being assembled. Both systems are portable, freely adjustable and provided with heavy bases. Photographic facilities include a 5 inch x 4 inch plate camera specially adapted for daylight use with the Schlieren system along with a high speed drum camera and a cine camera oprating at framing speeds up to 20,000 frames per second.

2.2. Operational Problems

The gas dynamics and supersonic tunnel facilities have proved to be simple to operate, reliable and comparatively trouble-free. However, there have been two operational problems which may of interest, and they are described below.

- Oil mist carry over: A slight oil deposition occurs on the tunnel test section windows after several runs indicating that the oil filtration is not completely effective. It is likely that this results from the fact that the oil mist filters have their highest removal efficiency for aerosol having particulate sizes down to 0.5 micron. It is possible that a very fine oil aerosol is passing through the filters with particulates in the size range 0.1 to 0.001

micron. A solution to the problem would be the use of a hydrocarbon selective absorbent which has a high removal efficiency in the size range where Brownonian motion predominates. For this reason, a set of activated carbon filters was manufactured and installed.

- Water Condensation in Pipelines: Operation of the gas dynamics facility without the air dryer is currently necessary for certain types of tests. As a result, liquid water condenses in the pipelines and pressure vessels and ultimately results in small rust flakes periodically discharging into a stagnation chamber. The solution to this problem appears to be use only dry air for all operations. The existing single dryer tower does not have sufficient drying capacity for some rig work where run durations of up to eight hours are required. Consequently, installation of the second dryer tower appears to be necessary, although much of this type of running can be done at higher outlet dew points than the supersonic tunnel operation, thus permitting longer run times before regeneration. It also appears desirable to introduce some form of direct reading humidity meter which could also be used to operate a warning system. As an interim measure to combat water collecting in the pipelines and fittings, all low points of the system have been provided with water drain valves.

3. CONCLUSION

A compressed air plant, providing 1.5 lb/sec of dry air at 100 to 115 psig and having a storage capacity of 1000 cubic feet, has been engineered and built. The system has been operating satisfactorily apart from an oil mist problem for which corrective measures are being investigated.

A supersonic blowdown wind tunnel, using air from the compressed air plant and exhausting to atmosphere, has also been built to simple, conventional design principles. Nozzle blocks for Mach 3 and parallel duct supersonic diffuser has been installed. Although stagnation pressure control is manual, the tunnel is designed for operation by only one man. Run time varies from 20 to 60 seconds and test section Reynolds numbers of about 10^6 per inch may be obtained.

These facilities along with subsonic wind tunnel facilities form the basis of aerodynamic, research and development works at the University of New South Wales [31-89]

4. ACKNOWLEDGEMENTS

The Author wishes to gratefully acknowledge the hard works and dedication of Barry Motson and the late Associate Professor Archer in the Design of the Gas Dynamics facility and the Supersonic Wind Tunnel.

REFERENCES

1. P. G. Pugh, L. C. Ward, 1966 Notes on the Automatic Control of a Blowdown Wind Tunnel' NPL Aero Report 1215
2. K. G. Conolan, 1964 Control equipment for the ARL Hypersonic Wind Tunnel', ARL TM 196
3. A. Pope, K. L. Goin, 1984 High speed wind tunnel testing', Wiley, NY
4. J. C. Gibbings, 1965 A Note on the Combination of a Contraction with a Supersonic Nozzle for Wind Tunnel', RAE TR 65066
5. A. Mc Cabe, 1964 Design of Supersonic Nozzle', ARC Fluid Motion Sub-Committee, ARC 25716, FM 3433, March
6. M. Sibulkin, 1956 Heat Transfer to an Incompressible Turbulent Boundary Layer and Estimation of Heat Transfer Coeffcients at Supersonic Nozzle Throats', JAS, 23 2 162
7. E. W. E. Rogers, B. M. Davis, 1956 A note on Turbulent Boundary Layer Allowances in Supersonic Nozzle Design ARC CP 333
8. K. N. Dodd, et al. 1964 Calculations for the Design of Nozzles' RAE TR 64021
9. M. Tucker, 1951 Approximate Calculation of Turbulent Boundary Layer Development in Compressible Flow NACA TN 2337

10. J. E. Beckwith, J. A. Moore, 1955 An Accurate and Rapid Method for the Design of Supersonic Nozzles NACA TN 3322

11. D. V. Faro, . (ed, 1964 Handbook of Supersonic Aerodynamics', NAVWEPS Rep 1488, 6 sec 17, Ducts, Nozzles and Diffusers',

12. D. Pierce, 1965 A Simple Flexible Supersonic Wind Tunnel Nozzle for the Rapid and Accurate Variation of Flow Mach Number RAE TR 65280

13. H. P. Liepman, 1955 Analytic Method for the Design of 2D Asymmeric Nozzles', J of Aero Sc., 701 709

14. M. Culley, 1966 The ARL Supersonic Propulsion Wind Tunnel: Redesign of the Nozzle Profile and preliminary Calibration', ARL ME 282

15. W. Daniels, 1956 Design and development of the North American Aviation Transonic Wind Tunnel', AGARD 67

16. J. Lukasiewicz, 1953 Diffusers for Supersonic Wind Tunnels', J of Aero Sci, 20 9 617 626

17. R. De Leo, J. Huerta, 1956 The Influence of Adjustable Diffuser Configurations on Minimum Required Starting and Operating Pressure Ratios for a Supersonic Wind Tunnel', Proc of 5th Biennial Tech Conf. Uni of Minnesota, Research Report 137

18. N. B. Tucker, 1956 Data on Temperature Stabilisation and Diffuser Performance of the 5 inch x 5 inch Pilot Blowdown Wind Tunnel, AGARD 92

19. J. Lucasiewicz, 1964 Pressure Measurement in Wind Tunnel S3, WRE TN HSA 132

20. Ferri, S. F. Bogdonoff, 1956 Design and Optimisation of Intermittent Supersonic Boundary layer and Estimation of Heat Transfer Coefficients at Supersonic Nozzle Throats', JAS 23 23 62

21. L. E. Leavy, 1956 A supersonic Wind Tunnel for Mach Numbers up to 3. AGARD 70

22. R. Herman, 1956 A Basic Theorem Concerning Pressure Recovery of Symmetric Supersonic Diffusers', Proc of 5th Biennial Tech Conf. Uni of Minnesota, Research Report 137

23. L. L. Beranek, 1954 Acoustics', McGraw-Hill, NY

24. Ahmed N. A. Design Features of the Low Speed 4 ft x 3 ft Return Circuit Wind Tunnel'

25. D. Callaway, L. G. Ramer, 1952 The use of Perforated Facings in Designing Low-Frequency Resonant Absorbers J of Acoustical Soc of America, 24 3 309 312

26. L. L. Beranek, S. Labate, U. Ingrad, 1955 Noise Control for NACA Supersonic Wind Tunnel J of Acoustical Soc of America, 27 1 85 98

27. E. Myer, F. Miechel, G. Kurtze, 1958 Experiments on the Influence of Flow on Sound Attenuation in Absorbing Ducts', J of Acoustical Soc of America, 30 3 165 174

28. U. Ingard, R. H. Bolt, 1951 Absorption Characteristics of Acoustical Material with Perforated Facing J of Acoustical Soc of America, 21 2 533

29. L. L. Beranek, 1960 Noise Reduction' McGraw Hill, NY,

30. A. Anderson, 1963 Flow Characteristics of a 12 Inch Intermittent Supersonic Tunnel', AEDC Tech Doc Report AEDC-TDR- 63 203

31. N. A. Ahmed, 2010 Wind driven Natural-Solar/Electric Hybrid Ventilators', in ' Wind Power ', Section D: The Environmental Issues", Chapter 21, edited by S. M. Muyeen Kitami, published by In-Tech Organization, Austria, 978-9-53761-981-7

32. J. Lienand, N. A. Ahmed, 2011 Wind driven ventilation for enhanced indoor air quality', invited Chapter, in 'Chemistry, Emission, Control, Radiaoactive Pollution and Indoor Air Quality', edited by Nicholas A Mazzeo, published by In-Tech Organization, Austria, 978-9-53307-570-9

33. N. Findanis, N. A. Ahmed, 2011 Wind tunnel 'concept of proof' investigations in the development of novel fluid mechanical methodologies and devices', invited Chapter, in '

Wind Tunnels and Experimental Fluid Dynamics Research', edited by J. C. Lerner and U. Boldes, published by In-Tech Organization, Austria, 978-9-53307-623-2

34. N. A. Ahmed, R. L. Elder, C. P. Foster, J. D. C. Jones, 1987 A Novel 3D Laser Anemometer for Boundary Layer Studies', ASME Conf, Boston, USA, 15th December. Also in the 3rd International Symposium on Laser Anemometry edited by A Dybs et al, ASME, The Fluids Engineering Division, 55 175 117

35. N. A. Ahmed, R. L. Elder, C. P. Foster, J. D. C. Jones, 1990 Miniature Laser Anemometer for 3D Measurements J of Measurement Sc Technol, 1 272 276

36. N. A. Ahmed, R. L. Elder, C. P. Foster, J. D. C. Jones, 1990 Miniature Laser Anemometer for 3D Measurements Engineering Optics, 3 2 191 196

37. N. A. Ahmed, R. L. Elder, C. P. Foster, J. D. C. Jones, 1991 Laser Anemometry in Turbomachines IMechE Proc, Part G, J of Aerospace Engineering, 205 1 12

38. N. A. Ahmed, S. Hamid, R. L. Elder, C. P. Foster, J. D. C. Jones, R. Tatum, 1992 Fibre Optic Laser Anemometry for Turbo machinery Applications', Optics and Lasers in Engineering, 15 2-3 193 205

39. N. A. Ahmed, R. L. Elder, 2000 Flow Behaviour in a High Speed Centrifugal Impeller Passage under Design and Off-design Operating Conditions', Fluids and Thermal Engineering, JSME International 43 1 22 28

40. R. G. Simpson, N. A. Ahmed, R. D. Archer, 2000 Improvement of a Wing Performance using Coanda Tip Jets', AIAA Journal of Aircraft, 37 1 183 184

41. A. Gatto, N. A. Ahmed, R. D. Archer, 2000 Investigation of the Upstream End Effect of the Flow Characteristics of a Yawed Circular Cylinder', The RAeS Aeronautical Journal, 104 1033 253 256 pp125-128

42. A. Gatto, N. A. Ahmed, R. D. Archer, 2000 Surface Roughness and Free stream Turbulence Effects on the Surface Pressure

over a Yawed Circular Cylinder', AIAA Journal of Aircraft, 38 9 1765 1767

43. N. A. Ahmed, R. D. Archer, 2001 Performance Improvement of a Bi-plane with Endplates', AIAA Journal of Aircraft, 38 2 398 400

44. A. Gatto, K. P. Byrne, N. A. Ahmed, R. D. Archer, 2001 Pressure Measurements over a Cylinder in Crossflow using Plastic Tubing', Experiments in Fluids, 30 1 43 46

45. N. A. Ahmed, R. D. Archer, 2001 Post-Stall Behaviour of A Wing under Externally Imposed Sound', AIAA Journal of Aircraft, 38 5 961 963

46. N. A. Ahmed, R. D. Archer, 2002 Testing of a Highly Loaded Horizonatal Axis Wind Turbines designed for Optimum Performance', International Journal of Renewable Energy, 25 4 613 618

47. R. G. Simpson, N. A. Ahmed, R. D. Archer, 2002 Near Field Study of Vortex Attenuation using Wing Tip Blowing', The Aeronautical Journal 102

48. N. A. Ahmed, J. Goonaratne, 2002 Lift augmentation of a low aspect ratio thick wing at a very low angle of incidence operating in ground effect", AIAA Journal of Aircraft, 39 2

49. N. A. Ahmed, 2002 Implementation of a momentum integral technique for total drag measurement', International Journal of Mechanical Engineering and Education, 30 4

50. A. Pissasale, N. A. Ahmed, 2002 Theoretical calibration of a five hole probe for highly three dimensional flow', International Journal of Measurement Science and Technology, 13 1100 1107

51. A. Pissasale, N. A. Ahmed, 2002 A novel method of extending the calibration range of five hole probe for highly three dimensional flows", Journal of Flow Measurement and Instrumentation, 13 1-2 23 30

52. N. A. Ahmed, D. J. Wagner, 2003 Vortex shedding and transition frequencies associated with flow around a circular cylinder", AIAA Journal 41 3 542 544

53. D. H. Rashid, N. A. Ahmed, R. D. Archer, 2003 Study of aerodynamic forces on rotating wind driven ventilator', Wind Engineering 27 1 63 72

54. N. A. Ahmed, 2003 An acoustic energy concept for the design of a flow meter', International Journal of Vibration and Acoustics, 8 1 52 58

55. A. Pissasale, N. A. Ahmed, 2003 Examining the effect of flow reversal on seven-hole probe measurements', AIAA Journal, 41 12 2460 2467

56. A. Pissasale, N. A. Ahmed, 2004 Development of a functional relationship between port pressures and flow properties for the calibration and application of multi-hole probes to highly three-dimensional flows', Experiments in Fluids, 36 3 422 436

57. N. A. Ahmed, 2004 Demonstration of the significance and limitations of thin airfoil theory in the study of aerodynamic characteristics of an airfoil", IJMEE, 34 2 271 282

58. N. A. Ahmed, 2006 Investigation of dominant frequencies in the transition Reynolds number range of flow around a circular cylinder Part I: Experimental study of the relation between vortex shedding and transition frequencies', Journal of CSME, 19 2 159 167

59. N. A. Ahmed, 2006 Investigation of dominant frequencies in the transition Reynolds number range of flow around a circular cylinder Part II: Theoretical determination of the relationship between vortex shedding and transition frequencies at different Reynolds numbers', Journal of CSME, 19 3 317 326

60. S. Shun, N. A. Ahmed, 2008 Utilizing wind and solar energy as power sources for a hybrid building ventilation device Renewable Energy 33 6June 1392 1397

61. N. Findanis, N. A. Ahmed, 2008 The interaction of an asymmetrical localised synthetic jet on a side supported sphere Journal of Fluids and Structures 24 7 1006 1020

62. M. Longmuir, N. A. Ahmed, 2009 Commercial Aircraft Exterior Cleaning Optimization AIAA, Journal of Aircraft 46 1 284 290

63. S. J. Lien, N. A. Ahmed, 2010 Numerical simulation of rooftop ventilator flow Building and Environment 45 1808 1815

64. S. J. Lien, N. A. Ahmed, 2011 Effect of inclined roof on the airflow associated with a wind driven turbine ventilator Energy and Buildings 43 358 365

65. J. Lien, N. A. Ahmed, 2011 An examination of the suitability of multi-hole pressure probe technique for skin friction measurement in turbulentflow', in press, Journal of Flow Measurement and Instrumentation 22 153 164

66. C. Wu, N. A. Ahmed, 2011 Numerical Study of Transient Aircraft Cabin Flowfield with Unsteady Air Supply AIAA Journal of Aircraft 48 6 1994 2002

67. N. Findanis, N. A. Ahmed, 2011 Three-dimensional flow reversal and wake characterisation of a sphere modified with active flow control using synthetic jet Advances and Applications in Fluid Mech, 9 1 17 76

68. G. Behfarshad, N. A. Ahmed, 2011 Vortex flow asymmetry of slender Delta Wings', International Review of Aerospace Engineering, 4 3 184 188

69. G. Behfarshad, N. A. Ahmed, 2011 Reynolds Stress Measurement Over Four Slender Delta Wings', International Review of Aerospace Engineering, 4 4 251 257

70. N. A. Ahmed, 2011 Detection of Separation bubble using spectral analysis of fluctuating surface pressure', International Review of Aerospace Engineering', 4 4

71. G. Behfarshad, N. A. Ahmed, 2011 Effect of unsteady and sinusoidally varying free stream on turbulent boundary layer separation Advances and Applications in Fluid Mechanics 10 2 79 98

72. G. Behfarshad, N. A. Ahmed, 2011 Experimental Investigations of Sideslip Effect on Four Slender Delta Wings', International Review of Aerospace Engineering, 4 4 189 197

73. N. A. Ahmed, J. R. Page, 2011 Real-time Simulation as a new tool in Future Advanced Aerospace Project Design and Manufacturing Processes Advanced Materials Research 317-319 2515 2519

74. N. A. Ahmed, J. R. Page, 2011 Developing and integrated approach to advanced aerospace project design in tertiary education Advanced Materials Research 317-319 2520 2529

75. H. Riazi, N. A. Ahmed, 2011 Numerical investigation of four orifice synthetic jet actuators, International Review of Aerospace Engineering', 4 5 272 276

76. S. Shun, N. A. Ahmed, 2011 Airfoil Separation Control using Multiple Orifice Air Jet Vortex Generators AIAA Journal of Aircraft 48 6 1994 2002

77. N. A. Ahmed, 2012 Engineering solutions towards cost effective sustainable environment and living' Journal of Energy and Power Engineering, 6 2 155 167

78. S. Shun and N.A. Ahmed (2012), 'Rapid Prototyping of Aerodynamics Research Models', Applied Mechanics and Materials vols. 217-219, pp 2016-2025,Trans Tech Publications, Switzerland

79. G. Behfarshad, N. A. Ahmed, 2012 Investigation of Newtonian liquid jets impacting on a moving smooth solid surface Advances and Applications in Fluid Mechanics 12 1

80. S. Shun and N.A. Ahmed (2012), 'Design of a Dynamic Stall Test Rig', Applied Mechanics and Materials Vols. 215-216, pp 785-795, Trans Tech Publications, Switzerland

81. N. A. Ahmed, 2012 Novel developments towards efficient and cost effective wind energy generation and utilization for sustainable environment', Renewable and Power Quality Journal, 0217-2038X, 10 PL4

82. Y.Y.Zheng, N.A.Ahmed and W.Zhang (2012), 'Feasibility Study of Heat Transfer with Fluidic Spike', International Review of Aerospace Engineering, vol. 5, no.2, pp 40-45.

83. Y.Y.Zheng, N.A.Ahmed and W.Zhang (2012), Impact Analysis of Varying Strength Counter-flow Jet Ejection on a Blunt

Shaped Body in A Supersonic Flow, Advances and Applications in Fluid Mechanics, vol 12, no.2, pp 119-129.

84. J. Yen, N. A. Ahmed, 2012 Enhancing Vertical Axis Wind Turbine Safety and Performance Using Synthetic Jets', (in Press), Journal of Wind and Industrial Engineering,

85. G. Matsoukas, N. A. Ahmed (2012), 'Investigation of Ionic Wind as a Means of Generating Propulsive Force', International Review of Aerospace Engineering, vol. 5, no.2, pp 35-39.

86. J. Yen, N. A. Ahmed, 2012 Synthetic Jets as a Boundary Vorticity Flux Control Tool', (in press) AIAA Journal,

87. C.Wu and N.A.Ahmed (2012), 'A Novel Mode of Air Supply for Aircraft Cabin Ventilation', *Building and Environment*, Vol. 56, pp. 47-56

88. T. G. Flynn, N. A. Ahmed, 2005 Investigation of Rotating Ventilator using Smoke Flow Visualisation and Hot-wire anemometer', Proc. of 5th Pacific Symposium on Flow Visualisation and Image Processing, 27 29 September, Whitsundays, Australia, Paper PSFVIP-5

89. J. Yen, N. A. Ahmed, 2012 Parametric Study of Dynamic Stall Flow Field with Synthetic Jet Actuation ASME Journal of Fluids Engineering 134 071106 071101

6
CHAPTER

Design Features of a Low Turbulence Return Circuit Subsonic Wind Tunnel Having Interchangeable Test Sections

N. A. Ahmed[1]

[1] *School Of Mechanical and Manufacturing Engineering, University of New South Wales, Sydney, NSW, Australia*

1. INTRODUCTION

Wind Tunnels have played and are continuing to play a significant role in providing controlled test facilities in the aerodynamic research and development [1-43, 122-178].The present chapter describes in detail, the design features of a subsonic return circuit wind tunnel that is currently in operation at the Aerodynamics Laboratory of the University of New South Wales. It can be considered to be a general purpose low speed tunnel with a sufficiently large contraction ratio. It has a number of removable turbulence reduction screens to achieve low turbulence level. It also has the provision of removable principal test section and three alternative test section arrangements located at various parts of the wind tunnel circuit. The wind tunnel can provide a wind speed in the range of 0-170 ft/sec at the lowest turbulence level. The top speed can be 200 ft/sec, if a higher turbulence level and spatial non-uniformities produced by omission of the screens can be tolerated.

FIGURE 1. Side View of the Subsonic Wind Tunnel of the University of New South Wales

Floor space limitations of approximately of 65 ft x 12 ft have meant that the tunnel be vertical in the vertical plane. From such consideration and ease of wind tunnel experiments, the test section was placed at the laboratory floor level and the return circuit above the test section. The upper structure of the laboratory roof was too flimsy and inaccessible for satisfactory location of the fan and drive in that area so that the fan and the drive had to be at the floor level. The fan is, therefore, placed downstream of the test section and first diffuser and upstream of the first cascade corner. This unconventional arrangement is not, however, without precedent; similar layout has been used in the N.B.S. 4.5 ft low turbulence wind tunnel and Wichita University 10 ft x 7 ft wind tunnels [44-46].

2. GENERAL CONSIDERATIONS

The configuration chosen presents several design advantages as well as disadvantages. These are detailed below:

Advantages:

1. Because the fan is located in a comparatively high speed portion of the tunnel, a favourable flow coefficient for a given tip speed may be more easily obtained, leading to high rotor efficiency

2. Except in the case of high lift or very bluff models, good inlet flow conditions to fan are obtained. This situation does not always occur in tunnels with the conventional fan location immediately after the second cascade corner. Maldistribution of flow may exist due to faulty turning vane performance or the need to pass the fan rotor drive shaft through the second cascade turning vanes. This, in turn, leads to reduced rotor performance and increased noise levels.

3. Flow disturbances created by the fan and its tail fairing in the conventional arrangement may adversely affect the performance of the main return circuit diffuser and hence the wind tunnel. The closed circuit type of diffuser is very sensitive to malfunctions in this diffuser [44,46-48]

4. The long flow return path between the fan and test section aids in achieving a low open tunnel turbulence level. This permits a reduction in the number of screens for certain types of test.

Disadvantages:

1. Since the fan is located in line of sight of the test section, care must be taken in the fan design to keep the noise level at the lowest possible value. Sound waves cause air motions which produce an effect similar to that of turbulence and this may place a lower limit on the tunnel turbulence level [44, 45, 49-51] In a tunnel with conventional fan location, the higher noise frequencies are partly attenuated by the two sets of turning vanes separating the fan and test section. Sound power transmitted to the test section from the fan may,

however, be reduced by the tunnel breather slot or the use of ducts with acoustic absorbent inserts [49].

2. Since, for reasons of safety, the fan must be observed in the design of the screen to prevent its causing a high energy loss.
3. Care must be exercised in the design of the fan prerotator blades (if fitted) to render them comparatively insensitive to flow changes caused by the presence of high lifting or bluff models in the test section. The contraction ratio was approximately 7:1, similar to one employed in the N.B.S. 4.5 ft tunnel.

Considerable difficulties had to be overcome in the erection of the tunnel components, none the least of which were the strengthening of the comparatively light floor and roof structures of the laboratory so as to absorb lifting and installation stresses. In its present configuration, the tunnel has an overall length of 67.5 ft, an overall height of 27.5 ft and an overall length of 11.5 ft, excluding interchangeable test sections. Various components of the wind tunnel were built over a period, and the overall work from the start of design to manufacture of various components to final installation took about five years to complete.

3. DESIGN OF VARIOUS COMPONENTS OF WIND TUNNEL

The detailed design of the tunnel components is described in the following sections of this report.

3.1. Test Section Design

The principal test section of 50 inch x 36 inch cross section has the normal value of its width to height, i.e., $\sqrt{2}:1$ [44]. Wall corrections are readily available for this configuration. The test section length of 9.75 ft is within the recommended range for general purpose work of 2.5 to 3 times the equivalent diameter (3.94 ft). Test section fillets, having a side of 5 inches are installed to prevent poor corner flow and accommodate the test section fluorescent lighting.

The test sections are tapered a total of 7/16 inch at the downstream and so as to compensate for the negative static pressure gradient associated with boundary layer thickness increase along the flow. This correction, which was found to be unattainable by tapering the test section fillets, as is sometimes recommended, is calculated to be approximately correct at a test section speed of 160 to 180 m/s. A filtered breather slot is located downstream of the test section.

When the original layout was developed, provisions were made to provide arrangements for removable test sections in various parts of the tunnel circuit. Four such test sections have been provided for. The possible configuration for each of the four is described below:

1. A principal test section having dimensions of cross section of 50 inch x 36 inch and 9.75 ft long and a speed range of 20 to 200 ft/sec.
2. A large test section can be inserted between the screen box/settling chamber assembly and the contraction, the latter being rolled back on a rail system after removal of the principal test section. This large test section is an octagon having maximum dimensions of 10 ft x 10 ft x 9.75 ft and a speed range of from 2 to 30 ft/sec. This test section is useful for a range of industrial aerodynamics tests.
3. An open jet test section, in conjunction with an appropriate removable collector, to be used if required, by removal of the principal test section.
4. A vertical test section which may be interposed in the tunnel circuit in place of the fourth diffuser. This test section permits testing in a vertical airstream and is of octagonal section having maximum dimensions of 5.1 ft x 5.1.ft and a speed range of from 10 to 100 ft/sec.

Of the above four, the first two have been constructed. The test sections were constructed of waterproof quality plywood of either ¾ inch or 1 inch thick, supported on angle from frames. Large viewing windows are provided from ½ inch and ¾ inch thick Perspex set in aluminium frames. The principal test section is provided with doors which open up one complete side over a length of 5 ft and extend two-thirds of the way across the top of the test section to improve accessibility. The tunnel floor is provided with a 3 ft diameter incidence change turntable mounted on a wire bearing race and controlled by a worm and piston drive. The principal test section is removed by means of an overhead travelling trolley and rail system. The large test section is traversed into position by means of a transverse floor rail system which aligns the walls and then by a set of translation tables which move the test section axially forward approximately 4 inches to close the pressure seal. Tapered dowel pins are used to secure accurate alignment of internal airline surfaces and over centre clamps are used to secure the vertical sections together.

3.2. Screen Settling Chamber Design

Wind tunnel screens are required to perform at least two functions, that is, to reduce the:

1. test section turbulence level, and
2. airstream spatial non-uniformities before entrance into the contraction and test section

3.2.1. Turbulence reduction

It has been shown experimentally by Schubauer et al [52] that no turbulence is shed by a screen if the Reynolds number based on the wire material is less than 30 to 60, the exact value depending upon the mesh size and wire diameter. Thus to obtain a low test section turbulence level, the turbulence reduction screens must be placed in a low speed region well upstream of the test section and contraction must consist of wires of the smallest diameter that are consistent with the strength required.

Batchelor [53] reports from experimental work that *'u'* and *'v'* turbulence components are reduced by factors of 0.36 and 0.54 respectively for wire screens having a resistance co-efficient of 2.0. According to additional experimental work by Dryden and Schubauer [B6), the mean turbulence intensity is reduced by the factor of 0.58 for k=2.0 screen and they propose the following relationship based on experiment but confirmed by appropriate theory:

$$U'_1/U'_3 = (1+k)^{-0.5}$$

U'1 and *U'3* are the mean turbulence intensities before and after the screens respectively. The relationship between the screen open area ratio or porosity and resistance co-efficient is best found from the data of Annand [54].

The analysis of Batchelor and Drydoen and Schubaureer reveal that it is best to employ a number of screens in series and that of Batchelor indicates that it is the reduction of 'v' component which is most difficult. Relation of the 'v' component to the required level will automatically ensure that the 'u' component is reduced to a correspondingly low value.

3.2.2. Flow non-uniformity reduction

The screens are also required to reduce the flow spatial non-uniformities before the airstream enters the contraction.

A theoretical analysis by Batchelor [55] and an earlier analysis by collar [56] have shown that for steady non-uniform flow, the *U* component non-uniformities are reduced in the ratio:

$$(2-K)/(2+k)$$

This expression implies that if k=2, the non-uniformities are completely removed. The analysis by Batchelor [53] indicates that the reduction factor can be more accurately expressed as:

$$(1-\alpha+\alpha K)/(1+\alpha+K)$$

where α is the screen deflection coefficient defined as the ratio of (air exit angle)/(air entry angle)

Taking the approximate value of α [56], the reduction factor for k=2 and α =0.64 is seen to be 0.1. Batchelor gives the theoretical reduction for V or transverse velocity non-uniformity component as α or 0.64 for a screen of resistance coefficient 2.0.

3.2.3. Limitations on screen arrangement

Batchelor analysis indicates that the 'v' component of turbulence is reduced by increasing k to a value of 4. However, screens having a resistance coefficient greater than 2 are not normally used, particularly for the final screen, for the following reasons:

1. Non-uniformity of weave of high resistance coefficient commercial screen materials produce flow disturbances which can have an adverse effect on test section flow distribution and turbulence level

2. Works by Bradshaw [57], Patel [58] and De Bray [59] have revealed that the final screen open-area ratios of less than 60% are likely to cause the development of flow instabilities of the type described by Morgan [60]. These instabilities produce small angular deviations in the flow downstream of the screens. De Bray suggests that a system of helical vortices originates at the screens and persists through the contraction and interacts with the test section boundary layers. The ultimate effect is to cause lateral variations in thickness and skin friction distribution in the test section boundary layers. Patel also reports that a similar effect is apparent if the screens are allowed to accumulate a build up of dust. Although a single screen resistance coefficient of 2.0 implies screen porosities of about 50%, it is necessary to use, at least for the final screen, a resistance coefficient of approximately 1.4 at 30 ft/sec in order to achieve a porosity of 57%. This is equivalent to a 20 mesh by 30 or 31 gauge wire screen.

There is also evidence to suggest that test section boundary layer disturbance of the type previously mentioned may be avoided by the use of a precision honeycomb located downstream of the last screen [B1,B9,B17]. However, such a device must have very small cell sizes, be of precision, and hence costly, construction and must be located in a very long settling length upstream of the contraction so as to reduce test section turbulence to a value equivalent to that obtained by the use of screen alone.

If screens are used, the attainment of a low turbulence level requires that use of several turbulence reduction screens each with a resistance co-efficient of less than 2. Following suggestion by Perry

[B10], it appears reasonable to optimise the screen configuration by the selection of individual screen resistance coefficients which give the maximum reduction in turbulence intensity and spatial non-uniformity with the minimum overall loss. However, in this tunnel, four screens of equal porosity give almost the optimum performance

3.2.4. Screen spacing and settling length

Because of space limitations, it is not usual in wind tunnel design to allow the full length between the turbulence reduction screens required for complete decay of the turbulence introduced by the screen wires. Dryden and Abbott [45] suggest that the turbulence is of the order of the wire diameter wire at a distance of about 200 wire diameters downstream of a screen. A survey of various designs [51] indicates that inter-screen settling lengths to wire diameter ratios of as little as 250 are used. Dryden and Schubauer [62] found that no measurable effect on the test section turbulence level of the N.B.S. 4 ½ ft tunnel was observed when the inter-screen spacing was varied from 2 to 28 inches. Bradshaw and Pankhurst [44] suggest a distance of 500 wire diameters.

The parallel length after the last screen should, however, be as long as possible, consistent with the space available. Most designs for low turbulence wind tunnels appear to have minimum values of about 2000 to 3000 wire diameters [51]. Work of Manton and Luxton [63] shows that the final period of turbulent decay is reached after a distance of approximately 700 wire spacings.

The University of New South Wales 4 ft x 3 ft wind tunnel has a provision for four removable turbulence reduction screens which have an inter-screen settling length of 400 wire diameters and a final settling length of 2000 wire diameters based on the use of 30 gauge wire gauge. A larger final settling length could not be achieved due to inadequate allowance for the screens and turning vanes in the original aerodynamic layout. However, a removable screen facility permits a considerable variety in screen settling length arrangements. The final screen was 20 mesh by 30 or 31 gauge wire and the remaining screens were the same to reduce turbulence and spatial non-uniformities with minimum overall pressure loss.

Because of the long return path between the fan and test section and the closeness of the vane spacing in the fourth cascade, the empty tunnel turbulence level was of the order of 0.2 to 0.3 %, falling to 0.08 to 0.1% with four screens fitted. The similar N.B.S. tunnel had had a turbulence level of 0.26% without screens, decreasing to 0.04% with six screens fitted.

The screen box of the University of New South Wales tunnel is manufactured from ¾ inch waterproof quality plywood reinforced by steel angle iron frames. The wire screens are clamped by bolting between removable pairs of 3 inch x 2 inch Oregon frames which are a neat sliding fit between pairs of similar fixed frames. The movable frames are supported on overhead tracks by sets of small ball-bearing wheels. Ample space has been provided around the edges of the screen box to install spring loaded screen tensioners, or individual frame air seals. The removable frames are provided with adjustable transverse stops and quick acting clamps so as to ensure their accurate and rigid alignment. The screen box door is sealed by a refrigeration type hollow rubber seal and is locked in position by means of eight swing over bolts and large hand wheels. Extensions of the screen sliding tracks are provided outside the screen box to enable the screens to be removed easily.

4. CONTRACTION DESIGN

A large contraction ratio is desirable for many reasons, some of which are:

1. A low air speed is obtained in the settling chamber thus permitting the installation of several low loss turbulence reduction screens without excessive power absorption
2. Because of the resulting low air speed in the settling chamber, turbulence generated in the last screen is lower for a given wire diameter
3. For a well designed contraction, the ratio of turbulence intensity to the mean speed will decrease as the mean speed increases at the test section entrance
4. A large contraction ratio, in conjunction with several damping screens, renders the tunnel test section characteristics least susceptible to disturbance in the tunnel circuit, such as those caused by high lift or bluff models [44].

In general, modern wind tunnels are designed for very low turbulence levels require contraction ratios of 12 to 16, in conjunction with up to six turbulence reduction screens. However, quite low turbulence levels may be obtained in wind tunnels with a contraction ratio of the order of 7:1, with four to six screens, and in conjunction with closely spaced vanes in the corner upstream of the settling chamber, as for example, in the N.B.S. 4 ½ ft tunnel [45].

The contraction ratio selected for the University of New South Wales tunnel produces reduction in the percentage longitudinal velocity non-uniformities by a factor of $1/n^2$ or 0.022 [B19] and of the mean RMS turbulence intensity by a factor of the order of [45] :

$$U'/U_T = \left[(2n/3 + 1/3n^2)^{0.5}\right]/n = 0.31$$

Taylor's alternative analysis suggests 0.4 to 0.8 [53]

There is as yet, no established exact design method for octagonal section wind tunnel contractions. Nevertheless, a design criterion common to all contraction is the avoidance of high wall curvature and large wall slope leading to possible adverse pressure gradients of strength sufficient to cause flow separation in either the contraction or test section.

This problem is particularly critical at the contraction entrance [43 and 46] and modern wind tunnels no longer use very small radius of curvature at the inlet end as was favoured before 1940 [44, 64 and 65]. It has been shown theoretically [66] that in order to obtain a uniform velocity distribution at exit, the velocity increase along the contraction must be monotonic but this condition is incompatible with the need for a finite contraction length. Most methods of design generally fall into of the five following categories:

1. Specification of an arbitrary contraction shape based on experience and/or the demands of the constructional material
2. A contraction shape given by the flow of a uniform stream about an arrangement of sources, sinks or vortex rings.
3. Specification of velocity distribution along the contraction axis leading to a derived contraction shape
4. Conformal transformation techniques

5. Specification of the contraction boundary velocity distribution in the hodograph plane and transformation to the x, r plane so as to derive the contraction shape in axisymmetric or two-dimensional flow.

Details of these methods can be found in References 64 to 84. The method employed for the University of New South Wales tunnel was to sketch in the shape, keeping in mind the demands of the constructional material techniques selected and the requirements for satisfactory performance [44, 46, 48, 64, 66, 69 and 85]. The contraction length was first estimated from the fact that, for contraction ratios of the order of 6 to 10:1, the ratio [51], the length to major inlet dimension, lies within 0.8 to 1.2.

The inlet and exit radii of curvatures are approximately 8 and 11 ft respectively for the University of New South Wales tunnel. The resultant contraction shape is very similar to that derived from an approximate theoretical solution by Cohen and Ritchie [64]. The contraction shape was approximately checked by the application of finite differences applied to the solution of the Laplace equation in radial symmetry [83]. A model was built and satisfactorily tested to confirm further the assumed design shape.

The contraction of the tunnel was manufactured from ¼ in marine ply, mitred and reinforced at the junction of the octagonal sides and built within accurately shaped frames of 3 inch x 2 inch Oregon. The Oregon frames were mounted at 1 ½ ft centres upon a base consisting of three longitudinal bearers of 6 inch x 4 inch Oregon. Flanged wheels and a rail system are mounted under the contraction to enable it to be moved axially along the tunnel centreline between the settling chamber and first diffuser.

5. DIFFUSER DESIGN

As mentioned in section 1, space limitation prevented the fitting of a controlled rapid expansion and the achievement of the optimum contraction ratio of 12 to 16:1. When it is possible to fit such an arrangement, a variety of flow stabilization methods of varying suitability are available for wide angle diffusers [86-94].

Considerable data is also available for the conventional diffuser design [98-104]. Unfortunately, however, little of this information has direct application to the design of three-dimensional octagonal section wind tunnel diffusers of any practical compact design must entail a certain amount of guess work or knowledge of previous experience in the selection of appropriate diffuser angles. For example, attempts to use the data of Ref D6 would indicate that for the large return diffuser of area ratio of 2.85:1, two-dimensional diffuser angles of up to 12^0 might be employed. However, experience with the square cross-section three-dimensional main return diffuser of the R.A.E. No. 2, 11 ½ ft x 8 ½ ft, wind tunnel1 indicated that equivalent cone angles of about 5^0 are satisfactory for this application. Shorter diffusers may employ somewhat larger angles and advantage has been taken of the fact in the design of the University of New South Wales tunnel where the equivalent cone angles used vary from 5.2^0 in the longest diffuser to a maximum of approximately 6 $¼^0$ in the shortest diffusers.

The first diffuser downstream of the test section is a particularly difficult design problem as the flow maldistribution caused by high lift and bluff models must be taken into account. Moreover, work by Willis [105] indicates that unsteady flow in the diffuser is responsible for a rise in a measured wall pressure spectra at low frequencies. The University of New South Wales tunnel has an essentially two-dimensional first diffuser with an included angle of 7 ¼ ^{0} and area ratio of 1.4:1 (equivalent cone angle of 3.4^0). Reference D6 indicates that a diffuser angle of up to 17^0 might be employed without separation for this diffuser.

Diffuser performance is also related to the inlet boundary layer thickness and free stream turbulence level [98, 99, 101-104). This makes the estimation of tunnel diffuser losses difficult. In the estimation shown in Table 1, the five diffusers contribute 37% of the tunnel loss, the first diffuser alone being about 14% of the tunnel loss. The design of the diffusion zone over the fan tail-fairing is a special problem and has been conveniently summarised by Russel and Wallis [106].

Diffuser numbers 4 and 5 of the University of New South Wales were built from ¾ inch thick exterior waterproof quality plywood with angle iron and 5 inch x 1 inch timber supporting frames. All sections are octagonal in shape as this permitted short length transitions to be made between the main components of the return circuit and circular fan ducting. The mitred sides of the octagons are constructed of 1/ inch ply mounted on 3 in x 2 in Oregon frames inside the main diffuser shell.

Diffuser No.1, the fan ducting and associated transitions are constructed from 16 gauge mild steel sheet which is reinforced with angle iron frames and rectangular bar steel frames and stringers.

Heavy Perspex windows and fluorescent lighting are fitted to enable easy visualisation of flow performance of the tunnel components. Each leg of the tunnel circuit between the turning vane cascades is provided with one or more quick opening doors for easy access. The doors are sealed with circular, foam rubber cord, formed into shape of an 'O' ring.

6. TURNING VANE DESIGN

It is well known that for abrupt rectangular corners, large aspect ratios and larges ratios of turning radius to inlet width are required to reduce the corner loss [107]. This has led to the post-second world war concept of closely spaced turning vanes to provide low loss, compact, wind tunnel corners.

In the past, it has been common to use thick profile aerofoil turning vanes because these can be designed to give air turning passages of approximately constant area, thus avoiding any expansion and possible flow separation around the passage between adjacent turning vanes. Such turning vanes are efficient in operation, but very difficult and expensive to construct. Winter [108] has shown that these thick vanes may be replaced by thin sheet metal turning vanes with little or no increase e in pressure loss at the corner. According to Winter[108], at a Reynolds number of 1.9×10^6 and for the same spacing to chord ratio (s/c) of 0.25, the thin sheet metal vanes reduced the vane loss to about 50% of that thick profiled turning vanes.

There is very little reliable information in the literature relating to turning vane losses for typical wind tunnel applications. The most extensive information is that reported by Salter [109] who obtained experimental data for both aerofoil profile and sheet metal circular arc turning vanes in the Reynolds number range of 6 x 10^4 to 1.9 x 1.9 x 10^5. It must be noted that the data presented by Salter does not employ the conventional cascade definition of spacing to chord (s/c) ratio in which the vane spacing is measured normal to the line joining the vane trailing edges. Salter defines a gap to chord ratio based on the distance or gap between the vane trailing edges measured normal to the parallel trailing edge tangents. It would appear that this data has been either misinterpreted or not adequately clarified in most of the subsequent literature [44]. Salter's data has been recalculated according to the conventional cascade definition of s/c ratio

The thin circular arc vanes tested by Salter appear to have a minimum loss co-efficient at an s/c ratio of between 0.3 and 0.4. The difference in the magnitude of the loss co-efficient for the Salter type 2 and 3 vanes could be due to the slightly different camber angles, but it is most likely due to the threefold increase in Reynolds number for the type 3 vanes. The series of tests by Ahmed revealed a considerable variation in loss coefficient with Reynolds number up to a value of about 4 x 10^5 after which the loss coefficient remained essentially constant. The curves designated Salter 2 and 3 are mean loss coefficients for a cascade corner including losses due to boundary layer and secondary flow effects. Salter also measured the loss coefficient for the potential flow region alone. The greater relative difference can be attributed to the fact that the lesser number of vanes and lower aspect ratio of the type 3 vanes contributes to a larger secondary flow loss. Salter concludes that for 90^0, thin circular arc turning vanes, having 10% straight tangent extensions on the leading and trailing edges, the mean loss coefficient should not exceed 0.1 for Reynolds numbers in excess of 2 x 10^5. Salter recommends that, to ensure flow stability, the gap chord ratio should be about 0.2 with a vane aspect ratio greater than 3. This gap chord ratio of 0.2 corresponds to an s/c ratio of 0.28 by the conventional cascade definition. Also evident from Salter's results is that the optimum s/c ratio for thick aerofoil profile vanes is in the region of 0.5 to 0.6.

The types of thin sheet metal vanes tested by Silberman[110] have a minimum loss coefficient at an s/c ratio of 0.5 to 0.7 depending upon the vane shape. The curves shown represent the loss coefficients in the potential flow region only. Silberman's results for thick vanes indicate a minima at an s/c value of 0.5.

Since s/c is not the only parameter determining the turning vane design for wind tunnels, a choice must be made of either vane spacing 's' or chord 'c'. This apparent variation possible in this choice is exemplified by the values for the fourth cascade corners of two successful wind tunnels of roughly comparable size and performance, i.e., the R.A.E. 4 ft x 3 ft and N.B.S. 4 ½ ft tunnels. For the R.A.E. tunnel, an s/c ratio of 0.26 was selected using thick profiled turning vanes of 30 inch chord. For the N.B.S. tunnel, the s/c ratio was 0.52 with a chord of 2 7/8 inches, employing thin sheet metal vanes. These two designs represent opposite limits of cascade performance. The R.A.E. vanes appear to have been designed for low loss, whereas those of the N.B.S tunnel were designed for low turbulence. The large chord of the R.A.E. vanes implies high Reynolds numbers and lower loss coefficients. In the N.B.S. tunnel1, the smaller blade spacings selected (approximately 1 ½ inches) resulted in a lower turbulence level measured at the screen location. The 'u' turbulence component of the N.B.S. tunnel1 referred to the settling chamber velocity and, measured in the settling chamber downstream of the fourth cascade, was about 2.3% and about 60% greater than the 'v' or 'w' components. This is a favourable design situation as it is the 'v' and 'w' components which are least reduced by passage through the screens and contraction. In the R.A.E. tunnel, the turbulence level in the comparable location was about 5 % and roughly equal for all three components.

It, therefore, appears that wind tunnel turning vanes can be constructed from thin sheet metal circular arcs, having an s/c ratio in the region of 0.28 to 0.35 and a passage aspect ratio of 6 or more. It appears that vanes for more than 90^0 corners should have a camber angle of 86^0 to 87^0 and that they should be set initially at a positive angle of about 3^0 to 4^0 with trailing edge angle of zero relative to the tunnel centreline at exit. The selection of the value of blade spacing depends upon the application envisioned. Low turbulence tunnels require that small blade spacing be used, for example, a spacing dimension of 2 inch or 3 inch would be unreasonable. Tunnels not requiring a low 'open tunnel' turbulence level might employ spacing dimensions of 12 to 24 inches. Additional compromises to be effected are those of cost and structural integrity.

Small vane spacings imply a large number of thin vanes of small chord with a resulting high cost and the possibility of vibration occurring due to relatively low vane natural frequency. Tunnels designed for low corner losses might be designed with a relatively large vane spacing and chord in order to ensure Reynolds numbers in excess of about 4×10^5. Salter suggests that a minimum of 20 turning vanes should be used in low loss corners.

The university of New South Wales tunnel employs s/c ratios of 0.25 and 0.27 for the first and the second cascade corners increasing to 0.31 for the third and fourth corners. Blade spacings vary from 2 to 5 inches and the number of turning vanes from 41 to 33 for the first and fourth cascade corners respectively. The maximum and minimum vane Reynolds numbers at design speed are approximately 5×10^5 and 2.4×10^5 for the first and fourth corners respectively. Turning vane t/c ratios vary between 0.7 to 1.5%.

Because the University of New South Wales wind tunnel cross section is octagonal at all cascade corners and the vane chord is an appreciable dimension, special care had to be taken in the design of the junction between the turning vanes and the octagonal fillet so as to prevent the airstream expanding and subsequently contracting in its passage around the junction zone. The problem was solved by the manufacture of special concave and convex cross sections which were fitted in the cascade corner fillets. The shape of these special corner sections was generated so as to provide a straight line intersection normal to the vane span at the junction of each turning vane and the corresponding corner fillet.

All turning vanes were produced from 10 gauge (1/8 inch) mild steel plate by brake pressing. The turning vanes are set in mild steel plate supporting frames which are reinforced with angle iron.

7. FAN AND DRIVE SYSTEM DESIGN

The fan must, by reason of its location downstream of the test section, pose certain design problems as outlined before. These relate to noise level and sensitivity to flow maldistribution caused by high lift or bluff models in the test section.

In general, the design methods of Wallis have been employed [111-113], together with additional experimental data [114-115]. A design utilising 100% pre-rotation has been developed in conjunction with N.P.L. type flow straighteners so as to ensure good efficiency over a wide range of flows together with reduced possibility of stall of the cascade corner vanes immediately downstream of the fan nacelle fairing.

The location of the fan in a relatively high speed portion of the tunnel is associated with a mean rotor blade flow co-efficient of 0.56, which approaches the optimum range of flow coefficients for high fan rotor efficiency with the amount of pre-rotation employed. However, there are conflicting fan duty requirements due to the need for relatively high pressure rise and low fan noise level.

As may be calculated from the estimated tunnel pressure loss characteristic, the fan duty required is 3.8 in w.g. pressure rise at a flow of 1f 122,000 CFM. The tunnel coefficient utilisation is:

(test section energy)/ (Σ circuit losses) = 1.6 to 2.3

depending on the number of screens used.

These requirements have led to the selection of an 8-bladed fan rotor of 5 ft diameter, limited to a maximum tip speed of 315 ft/sec. The rotor blade chords vary from 9.9 inches at the root to 6.4 inches at the tip.

The noise spectrum from an axial flow fan can be described as consisting of two components: 'broad band' noise and 'discrete frequency' noise.

Broad band noise is attributed to two basic mechanisms: vortex shedding from blade boundary layers and interactions between the blading and random turbulence in the intake flow. The theoretical analysis of Refs 116 and 117 show that, for rotor blades operating at their design point, the vortex shedding component of broad band noise is proportional to blade relative velocity to the power 5.6 and that the intake turbulence interaction component is proportional to relative velocity to the power of 4. Reduction in broad band noise can thus be realized mainly by keeping flow velocities adjacent to solid boundaries and, specifically, blade tip velocities, to minimum values consistent with satisfactory aerodynamic performance.

Discrete frequency noise is caused by periodic aerodynamic interaction between fixed and moving blade rows. Like broad band noise, discrete frequency noise has two basic mechanisms. These are the force fluctuations on individual blades which arise from variations in mean velocity of the incoming flow. The data in Refs 116-118 indicate that interaction noise is strongly dependent upon pre-rotator-rotor axial spacing and the shape and size of the individual pre-rotator vanes. The axial spacing affects mainly the potential pressure field interaction mechanism and the vane shape, the mean velocity variation mechanism. As an example, the discussion to Ref 118 indicates that the pressure variation due to the wake persistence is still about 10% of the maximum theoretically possible at a distance equal to one stator chord downstream, for typical accelerating cascades. Experimental data seems to indicate that, consistent with satisfactory aerodynamics, interaction noise is considerably reduced by using separations between stator an rotor of three-quarters to one vane chord in conjunction with small vane areas and slender profiles.

Blade sections chosen for the pre-rotators and rotor are C4 compressor sections on circular arc camber lines [112,114-115 and 119]. These sections give high isolated aerofoil lift coefficients at angles of incidence of 3^0 to 4^0 and have a high stalling incidence. The straightener design is based on the use of the symmetrical NACA 0012 section which starts to stall at about $\pm 14^0$ in the isolated aerofoil condition. Pre-rotator blades of cambered plates were considered [117] because of their comparatively low cost but were abandoned in view of their relatively poor performance under off-design conditions when compared with C4 sections.

Another parameter requiring careful selection was the choice of boss ratio as this affects the overall efficiency of the fan and tail fairing diffuser assembly. Due to the proximity of the first cascade corner, this ratio was fixed at a value of 0.4 which is less than optimum for the rotor alone.

The fan rotor blades have been stressed for centrifugal loading, torsional loads and loads due to non-coincident profile centroids and estimates have been made of the blade natural frequencies [120-121]. The fan rotor was dynamically balanced to an effective centre of gravity displacement of 3 to 5 microns.

The fan design requires a power output of 90 HP at 1200 RPM and a variety of fan drive schemes were considered. Thus a 90 HP compound wound DC motor and ancillaries that included switchgear and speed variation equipment were purchased. The Ward Leonard type speed control system proposed presented considerable difficulty in providing tunnel automatic dynamic head control. In addition, aerodynamic problems were encountered in designing the drive arrangement. A conventional shaft drive through the first cascade was at first envisaged but abandoned when it was realised that the required fairing through the cascade turning vanes caused severe blockage of a component which was already heavily loaded aerodynamically. A direct mechanical drive through a right angle bevel gearbox was next considered. However, a large fairing was needed for the drive shaft and problems were encountered in a gearbox design due to high power transmission requirements in a confined space. Alternative drive systems such as eddy-current variable speed couplings and Thyristor controlled DC drives were also investigated. All these units were costly and suffered from the same basic disadvantage that the prime movers, being large, had to be located outside the tunnel and required some sort of drive shaft arrangement through the tunnel structure to the fan rotor.

Thus the feasibility of using a hydraulic drive system was studied. This system comprises an axial piston hydraulic pump driving similar motor unit and is of the same order of cost as the other systems. The system has many advantages, the main ones being:

1. The drive motor is only 10 inches in diameter and 20 inches long for maximum power output of 125 HP at 1200 RPM. It fits radially inside the fan nacelle fairing where the local diameter is 23 to 24 inches. This eliminates aerodynamic problems associated with a drive shaft through the tunnel structure.
2. Automatic tunnel dynamic head control can be obtained with conventional pneumatic control equipment to a repeatability of ± 0.4 %.
3. The motor speed is fully variable from 0 to 14oo RPM by means of a diaphragm actuator and conventional pressure regulator.
4. The hydraulic pump can be driven by a standard 415 volt, 3-phase induction motor, for which installed electrical capacity was available.

The system finally selected consists of a 150 HP induction motor of 92% efficiency, driving a 'Lucas' PM 3000 series, seven axial piston hydraulic pump fitted with servo-control of the swashplate angle. The servo is operated by a standard 3-15 psi diaphragm actuator. The pump provides high pressure oil at approximately 2300 psi which is supplied to, and returned from the motor by 1 ½ inch outside diameter high pressure tubes through the fan straightener and supporting vanes. Oil flow is approximately 3500 GHP and the overall efficiency of the combined pump and rotor unit is of the order of 82 to 85 %, over the complete speed range. The system also includes ancillary equipment such as a 70 gallon oil reservoir, an oil cooler, boost pump and oil filtration equipment. The main disadvantage of the arrangement is high noise level from the rotor. Provision was, therefore, included in the design for reducing noise transmission of both hydraulic pump and motor.

The fan and drive system and first cascade corner are mechanically isolated from the rest of the tunnel structure, and the laboratory floor, so as to prevent the possibility of any vibrations being transmitted to the test section or instrumentation.

The fan rotor is mounted on an overhung bearing assembly supported off the front of the straightener vane assembly. The straightener vanes are manufactured from ¼ inch mild steel plate with radial and longitudinal plate stiffeners which both provide torsional rigidity and define the aerodynamic profile of the straighteners. The front and the rear of the straightener vanes are attached to heavy steel diaphragm plates at the hub. The front diaphragm plate supports a rigid bearing assembly which carries the overhung fan rotor. The rear diaphragm plate carries another diaphragm plate to which is bolted the hydraulic motor. A flexible coupling connects the very short fan rotor drive shaft and the motor output shaft between the front and rear diaphragm plates. Provision is made in the rotor bearing design to absorb the 400 lb rotor thrust loading. The five straightener vanes have bolted-on cast aluminium nose and tail pieces with the sides sheathed in 16 gauge aluminium sheet.

The fan rotor is of built up construction with blades being held in split root fixings which are in turn clamped between mild steel shroud plates. The rotor blades are high quality aluminium alloy castings with large cylindrical root attachments which enable the blades to be adjusted to any angle by releasing the rotor shroud plate clamping bolts.

The pre-rotator vanes are aluminium alloy castings and are clamped between the shroud plates at the roots to form a rigid prerotator drum assembly. The nacelle nose and tail fairings are spun from 16 gauge aluminium alloy sheet. The nose fairing is bolted on to the front of the pre-rotator drum and the tail fairing to the rear diaphragm plate carrying the hydraulic motor.

Estimations have been made of the tunnel air temperature rise due to power dissipation around the circuit. It was found that without any form of tunnel air exchange or heat exchanger, the air temperature rose by as much as 10 to 15^0 C above ambient after a period of operation of about 10 minutes at a speed of 150 ft/s in the principal test section. This may be doubled for long periods of operation at 200 ft/sec.

The tunnel control system is reasonably straight forward. Instrumentation comprises an optical tachometer, electric drive motor anemometer and pressure gauges for hydraulic system. Electrical interlocks are provided against loss of hydraulic boost pressure and inadvertent starting of the hydraulic system with the hydraulic motor set at the maximum speed condition. Possible fan blade failures are provided for by a fan vibration cut-out switch.

8. SAFETY NET DESIGN

For safe operation, a wind tunnel fan must have a suitable safety net located immediately upstream of it to prevent models, or tools, passing through the fan blades. The location of the fan in the University of new South Wales tunnel requires that the safety net be located in the relatively high speed portion of the tunnel circuit. This in turn, requires that considerable care is exercised in the aerodynamic design of the safety net.

It is not unusual to find the safety net located before the first cascade corner even in tunnels with conventional fan layout. It is also known that such safety nets can result in considerable tunnel power expenditure. It was found during experiments on the pressure losses in the ARL 9 ft x 7 ft tunnel that the safety screen which was located before the first corner, contributed 28 % to the total losses. This was the largest of any component. However, the safety net used in this case was relatively coarse, interlocked and 'cylcone' wire mesh.

The University of New South Wales tunnel safety screen is conical in shape and inclined at 45^0 to the free stream direction in order to reduce the velocity component normal to the screen. This configuration also ensures that any object stopped by the screen will be forced to the outside against the tunnel walls. The screen is constructed specially from fine gauge stainless steel wire so as to ensure a low pressure loss. One end of the screen is rigidly held whilst the other end is supported on an energy absorbing spring support.

9. CONCLUSION

A general purpose return circuit low speed wind tunnel has been designed for the Aerodynamics Laboratory of the University of New South Wales. A contraction ratio of 7:1 and four turbulence reduction screens are used. Low turbulence level is achieved with the assistance of some innovative design features. The fan is located upstream of the first corner. Corner cascade and screen configurations have received special attention.

Other unusual aspects of the design are three sizes of interchangeable test sections in the speed ranges of 0-25 ft/sec, 0-100 ft/sec and 0 -200 ft /sec.

The fan is driven by a hydraulic motor which considerably simplifies power transmission and control problems in this application.

10. ACKNOWLEDGEMENTS

The Author wishes to gratefully acknowledge the hard works and dedication of Barry Motson and the late Associate Professor Archer in the Design of this Wind Tunnel

References

1. N Findanis, and N. A Ahmed, Wind tunnel 'concept of proof' investigations in the development of novel fluid mechanical methodologies and devices', invited Chapter, in 'Wind Tunnels and Experimental Fluid Dynamics Research', edited by J.C. Lerner and U.Boldes, published by In-Tech Organization, Austria, 978-9-53307-623-2July, 2011

2. N. A Ahmed, Wind driven Natural-Solar/Electric Hybrid Ventilators', in 'Wind Power', Section D: The Environmental Issues", Chapter 21, edited by S.M.Muyeen Kitami, published by In-Tech Organization, Austria, 978-9-53761-981-7February, 2010

3. J Lien, and N. A Ahmed, Wind driven ventilation for enhanced indoor air quality', invited Chapter, in 'Chemistry, Emission, Control, Radiaoactive Pollution and Indoor Air Quality', edited by Nicholas A Mazzeo, published by In-Tech Organization, Austria, 978-9-53307-570-9June, 2011

4. N. A Ahmed, R. L Elder, C. P Foster, and J. D. C Jones, A Novel, D Laser Anemometer for Boundary Layer Studies', ASME Conf, Boston, USA, 15th December 1987Also in the 3rd International Symposium on Laser Anemometry, edited by A Dybs et al, ASME, The Fluids Engineering Division, 55175117

5. N. A Ahmed, R. L Elder, C. P Foster, and J. D. C Jones, Miniature Laser Anemometer for 3D Measurements', J of Measurement Sc Technol, 11990272276

6. N. A Ahmed, R. L Elder, C. P Foster, and J. D. C Jones, Miniature Laser Anemometer for 3D Measurements', Engineering Optics, 321990191196

7. N. A Ahmed, R. L Elder, C. P Foster, and J. D. C Jones, Laser Anemometry in Turbomachines', IMechE Proc, Part G, J of Aerospace Engineering, 2051991112

8. N. A Ahmed, S Hamid, R. L Elder, C. P Foster, J. D. C Jones, and R Tatum, Fibre Optic Laser Anemometry for Turbo machinery Applications', Optics and Lasers in Engineering, 15nos 2 and 3, 1992193205

9. N. A Ahmed, and R. L Elder, Flow Behaviour in a High Speed Centrifugal Impeller Passage under Design and Off-design Operating Conditions', Fluids and Thermal Engineering, JSME International series B, 431February, 20002228

10. R. G Simpson, N. A Ahmed, and R. D Archer, Improvement of a Wing Performance using Coanda Tip Jets', AIAA Journal of Aircraft, 3712000183184

11. A Gatto, N. A Ahmed, and R. D Archer, Investigation of the Upstream End Effect of the Flow Characteristics of a Yawed Circular Cylinder',The RAeS Aeronautical Journal, 1041033253256March, 2000pp125-128

12. A Gatto, N. A Ahmed, and R. D Archer, Surface Roughness and Free stream Turbulence Effects on the Surface Pressure over a Yawed Circular Cylinder', AIAA Journal of Aircraft, 389September, 200017651767

13. N. A Ahmed, and R. D Archer, Performance Improvement of a Bi-plane with Endplates', AIAA Journal of Aircraft, 382398400March-April, 2001

14. A Gatto, K. P Byrne, N. A Ahmed, and R. D Archer, Pressure Measurements over a Cylinder in Crossflow using Plastic Tubing', Experiments in Fluids, 30143462001

15. N. A Ahmed, and R. D Archer, Post-Stall Behaviour of A Wing under Externally Imposed Sound', AIAA Journal of Aircraft, 385961963September-October, 2001

16. N. A Ahmed, and R. D Archer, Testing of a Highly Loaded Horizonatal Axis Wind Turbines designed for Optimum Performance', International Journal of Renewable Energy, 254613618January, 2002

17. R. G Simpson, N. A Ahmed, and R. D Archer, Near Field Study of Vortex Attenuation using Wing Tip Blowing', The Aeronautical Journal, 102March, 2002

18. N. A Ahmed, and J Goonaratne, Lift augmentation of a low aspect ratio thick wing at a very low angle of incidence operating in ground effect", AIAA Journal of Aircraft, 392March-April 2002

19. N. A Ahmed, Implementation of a momentum integral technique for total drag measurement', International Journal of Mechanical Engineering and Education, 3042002

20. A Pissasale, and N. A Ahmed, Theoretical calibration of a five hole probe for highly three dimensional flow', International Journal of Measurement Science and Technology, 13July, 200211001107

21. A Pissasale, and N. A Ahmed, A novel method of extending the calibration range of five hole probe for highly three dimensional flows", Journal of Flow Measurement and Instrumentation, 131-2March-April, 20022330

22. N. A Ahmed, and D. J Wagner, Vortex shedding and transition frequencies associated with flow around a circular cylinder", AIAA Journal, 413March, 2003542544

23. D. H Rashid, N. A Ahmed, and R. D Archer, Study of aerodynamic forces on rotating wind driven ventilator', Wind Engineering, 27120036372

24. N. A Ahmed, An acoustic energy concept for the design of a flow meter', International Journal of Vibration and Acoustics, 81March 20035258

25. A Pissasale, and N. A Ahmed, Examining the effect of flow reversal on seven-hole probe measurements', AIAA Journal, 4112200324602467

26. A Pissasale, and N. A Ahmed, Development of a functional relationship between port pressures and flow properties for the calibration and application of multi-hole probes to highly three-dimensional flows', Experiments in Fluids, March, 363March, 2004422436

27. S Shun, and N. A Ahmed, Utilizing wind and solar energy as power sources for a hybrid building ventilation device', Renewable Energy, 336June 200813921397

28. N Findanis, and N. A Ahmed, The interaction of an asymmetrical localised synthetic jet on a side supported sphere', Journal of Fluids and Structures, 247200810061020

29. M Longmuir, and N. A Ahmed, Commercial Aircraft Exterior Cleaning Optimization', AIAA, Journal of Aircraft, 461Jan-Feb issue, 2009284290

30. S. J Lien, and N. A Ahmed, 2010Numerical simulation of rooftop ventilator flow. Building and Environment, 4518081815

31. S. J Lien, and N. A Ahmed, 2011Effect of inclined roof on the airflow associated with a wind driven turbine ventilator. Energy and Buildings, 43358365

32. J Lien, and N. A Ahmed, An examination of the suitability of multi-hole pressure probe technique for skin friction measurement in turbulent flow', in press, Journal of Flow Measurement and Instrumentation, 222011153164

33. C Wu, and N. A Ahmed, Numerical Study of Transient Aircraft Cabin Flowfield with Unsteady Air Supply, AIAA Journal of Aircraft, 486Nov-Dec issue, 201119942002

34. N Findanis, and N. A Ahmed, Three-dimensional flow reversal and wake characterisation of a sphere modified with active flow control using synthetic jet', Advances and Applications in Fluid Mech, 9120111776

35. G Behfarshad, and N. A Ahmed, Effect of unsteady and sinusoidally varying free stream on turbulent boundary layer separation', Advances and Applications in Fluid Mechanics, 10220117998

36. S Shun, and N. A Ahmed, Airfoil Separation Control using Multiple Orifice Air Jet Vortex Generators', AIAA Journal of Aircraft, 486Nov-Dec issue, 201119942002

37. N. A Ahmed, Engineering solutions towards cost effective sustainable environment and living' Journal of Energy and Power Engineering, 62February 2012155167

38. G Behfarshad, and N. A Ahmed, Investigation of Newtonian liquid jets impacting on a moving smooth solid surface', Advances and Applications in Fluid Mechanics, 1212012

39. N. A Ahmed, Novel developments towards efficient and cost effective wind energy generation and utilization for

sustainable environment', Renewable and Power Quality Journal, 0217-2038X, 10April issue, PL42012

40. Y. Y Zheng, N. A Ahmed, and W Zhang, Impact Analysis of Varying Strength Counter-flow Jet Ejection on a Blunt Shaped Body in A Supersonic Flow, (in press) Advances and Applications in Fluid Mechanics,(2012), 12(2), 119-129

41. J Yen, and N. A Ahmed, Parametric Study of Dynamic Stall Flow Field with Synthetic Jet Actuation', Journal of Fluids Engineering, Transactions of the ASME, (2012), 134 (7), 2012,45-53.

42. J Yen, and N. A Ahmed, Synthetic Jets as a Boundary Vorticity Flux Control Tool', AIAA Journal, (2013), 51(2),510-513

43. C Wu, and N. A Ahmed, A Novel Mode of Air Supply for Aircraft Cabin Ventilation', Building and Environment,(2012), 56, 47-56

44. P Bradshaw, and R. C Pankhurst, The Design of Low Speed Wind Tunnels', NPL Aero Report 1039, 1962

45. H. L Dryden, and I. H Abbott, The Design of Low Turbulence Wind Tunnels', NACA Report 940, 1949

46. K Razak, The University of Wichita 7 ft x 10 ft Wind Tunnel', University of Wichita Engg Report 022, April, 1950

47. H. B Squire, and K. G Winter, The RAE 4 ft x 3 ft Experimental Low Turbulence Wind Tunnel, Part I: General Flow Characteristics', ARC R&M 2690, February, 1948

48. MacPhailD.C., et el., 'The 2ft x 8 ½ ft Wind Tunnel at RAE, Farnborough', ARC R & M 2424, August, 1945

49. H Schun, The RAE 4 ft x 3 ft Experimental Low Turbulence Wind Tunnel, Part IV: Further Turbulence Measurements', ARC R&M 3261, June, 1953

50. P Bradshaw, Measurements of Free Stream Turbulence in some Low Speed Tunnels at NPL', R & M 3317, 1962

51. D. H Ferris, Measurements of Free Stream Turbulence in the RAE Bedford 13 ft x 9 ft Wind Tunnel', RAE Aero Report 1066, July, 1963

52. G. B Schubauer, et el 'Aerodynamic Characteristics of Damping Screens', NACA TN 2001

53. G. K Bachelor, Homogenous Turbulence', Cambridge University Press, 1953

54. P Annand, The Resistance to Air Flow of Wire Gauges', J Royal Aero Soc, 141146March, 1953

55. G. K Bachelor, On the Concept and Properties of the Idealised Hydrodynamic Resistance', ACA Report ACA-13, 1945

56. A. R Collar, The Effect of a Gauge on Velocity Distribution in a Uniform Duct', ARC R&M 1867, 1939

57. P Bradshaw, The Effect of Wind Tunnel Screens on 2 D Boundary Layers', NPL Aero Report 1085, December, 1963

58. N. P Patel, The Effects of Wind Tunnel Screens and Honey Combs on the Spanwise Variation and Honeycombs on the Spanwise Variation of Skin Friction in 2D Turbulent Boundary Layers', McGill University Mech Engg Tech Note 647October, 1964

59. B. G De Baray, Some Investigations into the Spanwise Non-uniformity of nominally 2D Incompressible Boundary Layers Downstream of Gauge Screens', ARC Fluid Memo, FM 3863, ARC 29271, July 1967

60. P. G Morgan, The Stability of Flow Through Porous Screens', J of Roy Aero Soc, 359362June 1960

61. J. L Lumley, Passage of a Turbulent Stream Through Honeycomb of Large Length-to Diameter Ratio', Tran ASME, Series D, 218220June 1964

62. H. L Dryden, and G. B Schubauer, The Use of Damping Screens for the Reduction of Wind Tunnel Turbulence', J of Aero Sci., 221228April 1947

63. M. J Manton, and R. E Luxton, Note on the Decay of Isentropic Turbulence', Inst of Engineers Austr. Conf. On Hydraulics and Fluid Mechanics', 9397November, 1968

64. M. J Cohen, and N. J. B Ritche, Low Speed 3D Contraction Design', J of Roy Aero Soc, 66232236April, 1962

65. G. K Batchelor, and F. S Shaw, A Consideration of the Design of Wind Tunnel Contractions', Aust Council Aeronautics Report, ACA-4, 1944

66. I. A Cohen, An Experimental Comparison of the Flow Induced in the Working Section of a Wind Tunnel by Contractions having 2D and 3D Flow Characteristics', Monash Univ Report, MME/65/1, August, 1965

67. B Thwaites, On the Design of Contractions for Wind Tunnels', ARC R & M 2278, March, 1946

68. F Cheers, Notes on Wind Tunnel Contractions', R & M 2137, March, 1945

69. J. C Gibbins, and J. R Dixon, D Contracting Duct Flow', Quarterly J of Mech and Applied Maths', 1024411957

70. J. D Stanitz, Design of 2D Channels with prescribed Velocity Distribution along the Channel Walls', Part I: Relaxation Solutions, NACA TN 2593, 1952, Part II: Solutions by Green's Function, NACA TN 2595, 1952

71. A. M. O Smith, and J Pierce, Exact Solutions of the Newmann Problem: Calculation of Non-Circulatory Plane and Axially Symmetric Flows about or Within Arbitrary Boundaries', Douglas Aircraft Co. Inc Report ES26988, April 1958

72. J. C Gibbins, Design of an Annular Entry to a Circular Duct', Aero Qtly, 103611959

73. P. A Libby, and H. R. M Reiss, The Design of 2D Contraction Sections', QAM, vol IX, April 1951

74. J. D Jackson, A Description of some Wind Tunnel Contraction Design Data which has been obtained using the Ferranti Mercury High Speed Digital Computer', ARC Report 23, 1962

75. R Harrop, A Method of Designing Wind Tunnel Contactions', J of Roy Aero Soc., 551691801951

76. R. H Smith, and C. T Wang, Contracting Cones giving uniform Throat Speeds', J of Aero Sc, 11October, 1944

77. R Jordinson, Design of Wind Tunnel Contractions', Aircraft Engg, 33294297October, 1944

78. L. G Whitehead, L. Y Wu, and M. H. L Waters, Contracting Ducts of Finite Length', Aero Qtly, 2254271February 1951

79. H. S Tsien, On the Design of the Contraction Cone for a Wind Tunnel', J of Aero Sci., 1068701943

80. B Szczeniowski, Contraction Cone for a Wind Tunnel', J of Aero Sci., 103113131943

81. M. J Lighthill, A new Method of 2D Aerodynamic design', ARC R & M., 2112, 1945

82. G. M Lilley, Some Theoretical Aspects of Nozzle Design', M.Sc. Thesis, UL, 1945

83. L. C Woods, A new Relaxation treatment of Flow with Axial Symmetry', Qtly J of Mech and Appl Maths, 4Part 3, 3583701951

84. W. T. F Lau, An Analytical Method for the Design of 2D Contactions', J of Roy Aero Soc. 68January, 1964

85. C Salter, and W. C Raymer, The NPL 7 ft Wind Tunnel: Sundry Notes and Comments following Measurements of Flow Distribution, Wall Pressures etc.,' NPL Aero Note 1023, October 1963

86. G. B Schubauer, and W. G Spangenberg, Effect of Screens in Wide Angle Diffusers', NACA Report 949, 1949

87. H. B Squire, and h Hogg, Diffuser Resistance Combinations in Relation to Wind Tunnel Design', RAE Report Aero 1933, ARC Report &628, April 1944

88. J. B Feir, The Effects of an Arrangement of Vortex Generators Installed to Eliminate Wind Tunnel Diffuser Separation', UTIAS TN 87june 1965

89. O. G Feil, Vane Systems for Very Wide Angle Subsonic Diffusers', Trans ASME Series D, J of Basic Engg, 759764December, 1964

90. C. A Moore, and S. J Kline, Some Effects of Vanes and of Turbulence in 2D Wide Angle Subsonic Diffusers', NACA TN 4080, June 1958

91. D. L Cochran, and S. J Kline, Use of Short Flat Vanes for Producing Efficient Wide Angle Subsonic Diffusers', NACA TN 4309, June 1958

92. R. W Fox, and S. J Kline, Flow Regime Data and Design Methods for Curved Subsonic Diffusers', Tans ASME, Series D, 843023161962

93. S. J Kline, C. A Moore, and D. L Cochran, Wide Angle Subsonic Diffusers of High Performance and Diffuser Flow Mechanisms', J of Aero Sci, 246469470June 1957

94. V. K Migay, Investigations of Finned Diffusers', USAF Systems Command,FTD-TT-6386October, 1962

95. R. R Wollett, Preliminary Investigation of Short 2D Subsonic Diffusers', NANA RM E56C02, May, 1956

96. K. N Raju, and D. M Rao, Experiments on the Use of Screens and Splitters for Flow Control in Wide Angle Diffuser', Indian N.A.L., TN-AE-2464April 1964

97. D. M Rao, and K. N Raju, Experiments on the use of Splitters for Control in Wide Angle Diffusers', Indian N.A.L. TN-AE-2664December 1964

98. J. R Henry, et alSummary of Subsonic Diffusers', NACA L56F05, October 1956

99. D. J Cockrell, and E Markland, A Review of Incompressible Diffuser Flow', Aircraft Engg, 286292October, 1963

100. G. N Patterson, Modern Diffuser Design', Aircraft Engg, 267273September, 1938

101. D. J Cockrell, and E Markland, The Effects of Inlet Conditions on Incompressible Fluid Flow in Conical Diffusers', Int J of Mech Sci, 81251391966

102. S. J Kline, D. E Abbott, and R. W Fox, Optimum Design of Straight walled Diffusers', Tans ASME J of Basic Engg, Paper 58-A1959

103. F. A. L Winternitz, and W. J Ramsay, Effects of Inlet Boundary Layer on Pressure Recovery, Energy Conversion and Losses in Conicl Diffusers', J of Roy Aero Soc, 61116124February, 1957

104. B. A Waitman, R. L Reneau, and S. J Kline, Effect of Inlet Conditions on Performance of 2D Subsonic Diffusers', Trans ASME J of Basic Engg, Paper 60-WA1960

105. J. A. B Wills, Spurious Pressure Fluctuations in Wind Tunnels', NPL Aero Report 1237, July 1967

106. R. A Wallis, Axial Fans', Newnes, London, 1961

107. G. N Patterson, Corner Losses in Ducts', Aircraft Engg', 205208August 1937

108. K. G Winter, Comparative Tests of Thick and Thin Turning Vanes in the RAE 4 ft x 3 ft Wind Tunnel', ARC R & M 2589, August, 1947

109. C Salter, Experiments on Thin Turning Vanes', ARC R& M 2469, October, 1946

110. E Silberman, Importance of Secondary Flow in Guide Vane Bends', Tech Paper 14Series B, 3rd Mid Western Conf on Fluid Mechanics, Univ of Minnesota, 1953

111. R. A Wallis, A Rationalised Approach to Blade Element Design', Conf on Hydraulics and Fluid Mechanics, Inst of Engineers, Aust, 1968

112. R. A Wallis, Optimisation of Axial Flow Fan Design', Ins of Engineers, Aust, M & C Trans, May 1968

113. B. A Russel, and B. A Wallis, A note on Annular Diffuser Design', CSIRO Int Rep 63July 1969

114. N Ruglen, Low Speed Wind Tunnel Tests on a series of C4 section Aerofoils', ARC Note ARL/A 275, July 1966

115. H. G Edmunds, and W. J Bartlett, Section Data of C4 Aerofoils', Power Jets Memorandum M1090April, 1940

116. D. J Myles, et el., '' The Design of Axial Flow Fans, Part II: Blade geometry for the Rotor and the Stator', DSIR, NEL Report 181, April 1965

117. I. J Sharland, Sources of Noise in Axial Flow Fans', J Sound Vibrations, 133023221964

118. I. J Sharland, Intake Noise from Axial Flow Turbochargers and Compressors', Proc Inst of Mech Engineers', 182373771968

119. A. D. S Garter, Blade Profiles for Axial Flow Fans, Pumps' compressore', Proc I Mech E, 175157758061961

120. M. J Schilhans, Bending Frequency of a Rotating Cantilever Beam', Trans ASME J Applied Mech and Maths, 2830March 1958

121. RoxbeeCox, H., (ed) 'Gas Turbine principles and Practice', Chapter 12: Vibrations', G. Newnes, London, 1955

122. N.A.Ahmed, 'Investigation of dominant frequencies in the transition Reynolds number range of flow around a circular cylinder Part I: Experimental study of the relation between vortex shedding and transition frequencies', Journal of CSME, vol.19, No.2, 2006, pp159-167

123. N.A.Ahmed, 'Investigation of dominant frequencies in the transition Reynolds number range of flow around a circular cylinder Part II: Theoretical determination of the relationship between vortex shedding and transition frequencies at different Reynolds numbers', Journal of CSME, vol.19, No.3, 2006, pp 317-326

124. G.Behfarshad and N.A.Ahmed, 'Vortex flow asymmetry of slender Delta Wings', International Review of Aerospace Engineering, Vol.4, No.3, 2011, pp 184-188

125. G.Behfarshad and N.A.Ahmed, 'Reynolds Stress Measurement Over Four Slender Delta Wings', International Review of Aerospace Engineering, Vol.4, No.4, 2011, pp 251-257

126. N.A.Ahmed, 'Detection of Separation bubble using spectral analysis of fluctuating surface pressure', International review of Aerospace Engineering', International Review of Aerospace Engineering', vol.4, no. 4, June, 2011

127. G.Behfarshad and N.A.Ahmed, 'Experimental Investigations of Sideslip Effect on Four Slender Delta Wings', International Review of Aerospace Engineering, Vol.4, No.4, 2011, pp 189-197

128. N.A.Ahmed and J.R.Page, 'Real-time Simulation as a new tool in Future Advanced Aerospace Project Design and Manufacturing Processes', Advanced Materials Research, Vols. 317-319 , 2011, pp 2515-2519

129. N.A.Ahmed and J.R.Page, 'Developing and integrated approach to advanced aerospace project design in tertiary education', Advanced Materials Research, Vols. 317-319 , 2011. pp 2520-2529

130. H. Riazi, and N.A. Ahmed, 'Numerical investigation of four orifice synthetic jet actuators, International Review of Aerospace Engineering', Vol.4, No. 5, 2011, pp 272-276

131. S. Shun and N.A. Ahmed, 'Airfoil Separation Control using Multiple Orifice Air Jet Vortex Generators', AIAA Journal of Aircraft, vol 48, no.6, Nov-Dec issue, 2011, pp 1994-2002

132. N.A.Ahmed, 'Engineering solutions towards cost effective sustainable environment and living' Journal of Energy and Power Engineering, Vol 6, No.2, February 2012, pp155-167

133. S. Shun and N.A. Ahmed, 'Design of a Dynamic Stall Test Rig', Applied Mechanics and Materials Vols. 215-216 (2012) pp 785-795, © (2012) Trans Tech Publications, Switzerland

134. G.Behfarshad and N.A.Ahmed, 'Investigation of Newtonian liquid jets impacting on a moving smooth solid surface', Advances and Applications in Fluid Mechanics, vol.12, no.1, 2012

135. S. Shun and N.A. Ahmed, 'Rapid Prototyping of Aerodynamics Research Models', Applied Mechanics and Materials Vols. 217-219 (2012) pp 2016-2025, © (2012) Trans Tech Publications, Switzerland

136. N.A.Ahmed, 'Novel developments towards efficient and cost effective wind energy generation and utilization for sustainable environment', Renewable and Power Quality Journal, ISSN 2172-038X, No. 10, April issue, pp PL4, 2012

137. Y.Y.Zheng, N.A.Ahmed and W.Zhang, 'Feasibility Study of Heat Transfer with Fluidic Spike', International Review of Aerospace Engineering, vol. 5, no.2, 2012, pp 40-45.

138. N.A.Ahmed, 'New Horizons of Applications of the 21st Century Aerodynamic Concepts from Aerospace to Power Generation and Utilization', Procedia Engineering, Elsevier Publications, vol. 49, 2012, pp 338-347

139. S. Shun and N.A. Ahmed, "Wind Turbine Performance Improvements Using Active Flow Control Techniques", Procedia Engineering, Elsevier Publications, vol. 49, 2012, pp 83-91

140. Matsoukas, G., and Ahmed, N.A., 'Experimental Investigation of Employing Asymmetrical Electrodes in Propulsion of Vehicles', Procedia Engineering, Elsevier Publications, vol. 49, 2012, pp 247-253

141. G.Behfarshad and N.A.Ahmed, 'Splash measurement of Newtonian Liquid Jets Impacting on a Moving Solid Surface' International Conference on Mechanical Engineering and Mechatronics, Ottawa, August 16-18, 2012

142. Y.Y.Zheng, N.A.Ahmed and W.Zhang, 'Heat dissipation using minimum counter flow jet ejection during spacecraft re-entry', Procedia Engineering, Elsevier Publications, vol. 49, 2012, pp 271-279

143. Wu, C., and Ahmed, N.A., 'Application of Flow Control Technique for Indoor Ventilation' Procedia Engineering, Elsevier Publications, vol. 49, 2012, pp 135-141

144. Riazi, H., and Ahmed, N.A., 'Effect of the ratio of specific heats on a small scale solar Brayton cycle', Procedia Engineering, Elsevier Publications, vol. 49, 2012, pp 263-270

145. Yen. J., and Ahmed, N.A., 'Improving the Safety and Performance of Small-Scale Vertical Axis Wind Turbine', Procedia Engineering, Elsevier Publications, vol. 49, 2012, pp 99-106

146. Wongpanyathaworn, M., and Ahmed, N.A., 'Optimising louver locations to improve indoor thermal comfort based on natural ventilation', Procedia Engineering, Elsevier Publications, vol. 49, 2012, pp 169-178

147. Findanis, N., and N.A. Ahmed, 'Control and Management of Particulate Emissions using Improved Reverse Pulse-Jet

Cleaning Systems', Procedia Engineering, Elsevier Publications, vol. 49, 2012, pp 338-347

148. T.G.Flynn, G.Behfarshad and N.A. Ahmed, 'Development of a Wind Tunnel Test Facility to Simulate the Effect of Rain on Roof Ventilation Systems and Environmental Measuring Devices', Procedia Engineering, Elsevier Publications, vol. 49, 2012, pp 239-246

149. J. Lien and N.A. Ahmed, Numerical evaluation of wind driven ventilator for enhanced indoor air quality', Procedia Engineering, Elsevier Publications, vol. 49, 2012, pp 124-134

150. N.A. Ahmed, 'Diverse Applications of Active Flow Control', Commissioned for publication in 'Progress in Aerospace Sciences', a commissioned paper by invitation of Journal board, expected publication, 2013

151. J. Lien and N.A. Ahmed, 'Indoor Air Quality Measurement with the Installation of a Rooftop Turbine Ventilator', Journal of Environment Protection', Vol.3, No.11, November 2012, pp1498-1508

152. J.Yen and N.A.Ahmed, 'Enhancing Vertical Axis Wind Turbine Safety and Performance Using Synthetic Jets', Journal of Wind and Industrial Engineering, vol.114, 2013, pp12-17

153. I.H. Salmom and N.A.Ahmed, "Delaying Stall by Acoustic Excitation Using a Vibrating Film Wing Surface", AIAA 22nd Applied Conference and Exhibit, 16-19 August 2004, Providence, Rhode Island, AIAA Paper No. 2004-4962

154. N.A.Ahmed, "Turbulent Boundary Layer Analysis of Flow in a Rotating Radial Passage", 2nd BSME-ASME International Conference on Thermal Engineering, 2-4 January, 2004, Dhaka, pp 325-333

155. T.G.Flynn and N.A.Ahmed, 'Investigation of Rotating Ventilator using Smoke Flow Visualisation and Hot-wire anemometer', Proc. of 5th Pacific Symposium on Flow Visualisation and Image Processing, 27-29 September, 2005, Whitsundays, Australia, Paper No. PSFVIP-5-214

156. N.Findanis and N.A.Ahmed, 'Wake study of Flow over a sphere', 25th AIAA Applied Aerodynamics Conference, San Francisco, USA, 8-10 June, 2006, AIAA-2006-3855

157. S.J. Lien and N.A. Ahmed, 'Skin friction determination in turbulent boundary layers using multi-hole pressure probes'25th AIAA Applied Aerodynamics Conference, San Francisco, USA, 8-10 June, 2006, AIAA-2006-3659

158. N.Findanis and N.A.Ahmed, 'A Flow Study Over a Sphere with Localised Synthetic Jet', 12th Australian International Aerospace Congress/12th Australian Aeronautical Conference19-22 March 2007, Melbourne, Victoria Australia

159. S.J. Lien and N.A. Ahmed, 'A novel method for Skin friction determination using multi-hole pressure probes', 12th Australian International Aerospace Congress/12th Australian Aeronautical Conference19-22 March 2007, Melbourne, Victoria Australia

160. B. Ahn and N.A. Ahmed, 'Internal Acoustic excitation to enhance the airfoil performance at high Reynolds number', 14th International Conference on Sound and Vibration, 9-12 July, 2007, Cairns, Australia

161. N.A.Ahmed, 'The Study of Spectral Properties of a Separation Bubble on a Circular Cylinder under the effects of free stream)turbulence 4th BSME-ASME International Conference on Thermal Engineering, 27-29 December, 2008, Dhaka

162. N.Findanis and N.A.Ahmed 'Active Flow Control Over a Bluff Body Utilising Localised Synthetic Jet Technology , 13th Australian International Aerospace Congress/13th Australian Aeronautical Conference, 9 - 12 March 2009, at the Melbourne Convention Centre, Melbourne, Australia

163. J.Lien and N.A.Ahmed 'Prediction of Turbulent Flow Separation with Pressure Gradient , 13th Australian International Aerospace Congress, 13th Australian Aeronautical Conference, 9 - 12 March 2009, at the Convention Centre, Melbourne, Australia

164. N.A.Ahmed and J.R.Page, 'An Improved Approach for Future Aerospace Design process', 13th Australian International Aerospace Congress/13th Australian Aeronautical

Conference, 9 - 12 March 2009, at the Convention Centre, Melbourne, Australia

165. N.A.Ahmed, 'Engineering solutions towards cost effective sustainable environment and living', International Conference on Mechanical, Industrial and Energy Engineering, 22-24 December, 2010

166. T.J.Flynn and N.A.Ahmed, 'An Investigation of Pitot Tube and Multi Hole Pressure Probe Performance Using a Wet Weather Wind Tunnel Test Section'14th AIAC, Melbourne, 28 Feb-3rd March, 2011

167. C.Wu and N.A.Ahmed, 'Using Pulsed Jet of Fresh Air to Control CO2 Concentration in an Air Cabin', 14th AIAC, Melbourne, 28 Feb-3rd March, 2011

168. C.Wu and N.A.Ahmed, 'Aircraft cabin flow pattern under unsteady air supply', 29th AIAA Applied Aerodynamics Conference, Honolulu, Hawaii, USA. 27 June 2011

169. H. Riazi and N.A. Ahmed, 'Numerical investigation on two-orifice synthetic jet actuators of varying orifice spacing and diameters', 29th AIAA Applied Aerodynamics Conference, Honolulu, Hawaii, USA. 27 June 2011

170. G.Behfarshad and N.A.Ahmed, 'Turbulent Boundary Layer Separation in Adverse Pressure Gradient', Proceedings of the 23rd Canadian Congress of Applied Mechanics, 5-9 June, 2011, pp 557-560

171. N.A.Ahmed, 'Novel developments towards efficient and cost effective wind energy generation and utilization for sustainable environment', Plenary Paper, International Conference on Renewable Energies and Power Quality ICREPQ'12, Santiago de Compostela, Spain, 28-30 March, 2012

172. S. Shun and N.A. Ahmed, "Utilizing Vortex Acceleration to Improve the Efficiency of Air Jet Vortex Generators", 6th AIAA flow control conference, New Orleans, 25-28 June 2012

173. H. Riazi and N.A. Ahmed, 'Efficiency enhancement of a small scale closed solar thermal Brayton cycle by a combined simple organic Rankine cycle', Proceedings of the ASME 2012

International Mechanical Engineering Congress & Exposition, IMCE2012, November 9-15, 2012, Houston, Texas, USA

174. J. Olsen, J.R.Page and N.A. Ahmed, 'A Hybrid propulsion system for a light trainer aircraft', *15th Australian International Aerospace Congress*, 25-28 February 2013.

175. Y.Y.Zheng, N.A. Ahmed and W.Zhang, 'A Feasibility Study of Bow Shock Wave Heat Dissipation using Counter-Flow Jet Activation', *15th Australian International Aerospace Congress*, 25-28 February 2013.

176. S.Shun and N.A. Ahmed, 'Experimental Investigation of the Effects of Fluid Acceleration upon Air Jet Vortex Generator Performance', *15th Australian International Aerospace Congress*, 25-28 February 2013.

177. C.Wu and N.A. Ahmed, 'Vectoring of Wall bounded Planar Ventilation Jet with Synthetic Jet Actuator', *15th Australian International Aerospace Congress*, 25-28 February 2013.

178. J.Yen and N.A. Ahmed, "Parametric study of synthetic jet on dynamic stall flow field using Computational Fluid Dynamics", *15th Australian International Aerospace Congress*, 25-28 February 2013.

7

CHAPTER

Experimental Aeroelastic Models Design and Wind Tunnel Testing for Correlation with New Theory

Deman Tang and Earl H. Dowell *

Department of Mechanical Engineering and Materials Science, Duke University, Box 90300, Hudson Hall, Durham, NC 27708-0300, USA

ABSTRACT

Several examples of experimental model designs, wind tunnel tests and correlation with new theory are presented in this paper. The goal is not only to evaluate a new theory, new computational method or new aeroelastic phonomenon, but also to provide new insights into nonlinear aeroelastic phenomena, flutter, limit cycle oscillation (LCO) and gust response.

Keywords: flutter; limit cycle oscillation; freeplay nonlinearity; frequency lock-in; aeroelastic model

1. INTRODUCTION

A very important function of wind tunnel models and testing is to verify a new aeroelastic theory or a new computational method. For this purpose, experimental aeroelastic model design and manufacturing of scaled models, model ground vibration test and wind tunnel testing are essential to success. In the past 20 years, the Duke aeroelastic group has designed many aeroelastic models and conducted wind tunnel tests to evaluate new theoretical aeroelastic theories and new computations methods. They include:

1) A high aspect ratio wing model. Several correlation studies for flutter and limit cycle oscillation (LCO) [1], limit cycle hysteresis response [2], gust response for clamped and flexibly suspended models [3,4] and Flutter/LCO suppression [5] were performed.

2) Wing like plate models, delta wing-store, flapping flag, yawed plate and folding wing. The wing tunnel tests were used to evaluate the von Karman nonlinear plate theory, and a new nonlinear inextensible beam and plate theory and also some high fidelity computation methods. Based on these models several correlation studies for flutter/LCO [6,7,8,9,10,11,12,13,14,15,16] and gust response studies [17,18,19] were performed.

3) Airfoil section with control surface freeplay. The wing tunnel tests were used to evaluate new approaches for the freeplay nonlinear and gust responses including Duke computational codes using the Peter's finite state airload aerodynamic theory, harmonic balance method, and the time marching integration based on state space equations [20,21,22,23,24] and the ZAERO code [25] based on the computational fluid dynamic (CFD) theory conducted by ZONA Technology, Inc.

4) All-movable tail with freeplay model at the root [26,27] to similar horizontal tail in the actuating mechanism freeplay nonlinearity of aircraft. Based on this model a computational code has been developed and evaluated.

5) A free-to-roll fuselage flutter model [28]. From measured wind tunnel data, one evaluates the predicted symmetric and anti-symmetric flutter/LCO theory.

6) An experimental oscillating airfoil model at high angles of attack for measuring aerodynamic response. A frequency "Lock-in" phenomenon is found in buffeting flow and compared to the theoretical results [29]. Also, an experimental airfoil model with a portial-span control surface is conducted

to measure the flap response of the portial-span induced by the buffeting flow [30].

Several examples of model designs, wind tunnel tests and experimental/theoretical correlation studies with the theory are presented in this paper. The goal is not only to evaluate a new theory, new computational method or new aeroelactic phonomenon, but also to provide new insights into nonlinear aeroelastic phenomena, flutter/LCO and gust response.

2. EXPERIMENTAL MODELS FOR MEASURING FLUTTER/LIMIT CYCLE OSCILLATION (LCO) RESPONSE TO EVALUATE A NONLINEAR STRUCTURAL THEORY

Theoretical and experimental correlation study for the aeroelastic stability and response of an aircraft with a high-aspect-ratio wing and low-aspect-ratio wing have been studied for many years from subsonic to supersonic flow. Most investigators have used linear beam and plate theories to simplify the wing structural model. However, a geometric structural nonlinearity may arise from the coupling between elastic flap or vertical bending, chordwise or edgewise bending and torsion for very high aspect ratio wings typical of Uninhabited Air Vehicles (UAV) and also for the low aspect ratio wings. Here, two kinds of experimental wing models, a high altitude long endurance (HALE) model and a wing-like (plate) model are considered to evaluate the new nonlinear structural theories.

2.1. High Altitude Long Endurance Models (Nonlinear Beam Structural Theory and ONERA Aerodynamic Model)

An experimental high-aspect ratio wing aeroelastic model with a tip mass has been constructed and a wind tunnel test conducted to measure the static aeroelastic response, flutter and limit cycle oscillations. The goal is to assess experimentally the theoretical results using nonlinear structural and aerodynamic theories. The theoretical structural equations of motion are based on nonlinear beam theory by Hodges and Dowell [31] and the ONERA aerodynamic stall model [32].

The experimental model includes two parts: a high-aspect ratio wing with a slender body at the tip, and a root support mechanism. The wing is rectangular, untwisted and flexible in the flap, lag and torsional directions. The wing is constructed from a precision ground flat steel spar with mass uniformly distributed along the wing span. The spar is 45.72 cm in length, 1.27 cm in width and 0.127 cm in thickness. It is inserted tightly into the wing root mechanism. In order to reduce torsional stiffness, the spar has multiple thin flanges along the span. The flange width is 0.127 cm and it is 0.318 cm in deep. There are 2 × 33 flanges uniformly and symmetrically distributed along the wing span and center line of the spar. There are 18 pieces of NACA 0012 airfoil plate uniformly distributed along the span. The pieces of the airfoil plate are made of aluminum alloy with 0.254 cm thickness. A precision aerodynamic contour of the wing model is obtained. Each airfoil plate has a slot 1.27 cm in width and 0.127 cm in thickness at the symmetry line. The spar is inserted through these airfoil plate slots and they are permanently bonded together. Each space between two airfoil plates is filled with a light wood (bass) covering the entire chord and span which provides the aerodynamic contour of the wing. This wood provides a slight additional mass and a small addition to the bending and torsion stiffness.

A slender body is attached to the elastic axis of the wing tip. The slender body is an aluminum bar, 0.95 cm in diameter and 10.16 cm in length. A paraboloidal forebody and an aftbody with 1.14 cm length are fixed to two ends of the bar. The forebody and aftbody are made of brass.

The slender body is symmetrical and is designed to provide enough torsional inertia to reduce the natural torsional frequency sufficiently to induce flutter in the velocity range of the wind tunnel.

The root support mechanism is a socket that allows a change of the steady angle of attack at the root. The root socket is mounted to the mid point of side-wall of the wind tunnel

Figure 1a shows a photograph of the aeroelastic wing model in the wind tunnel.

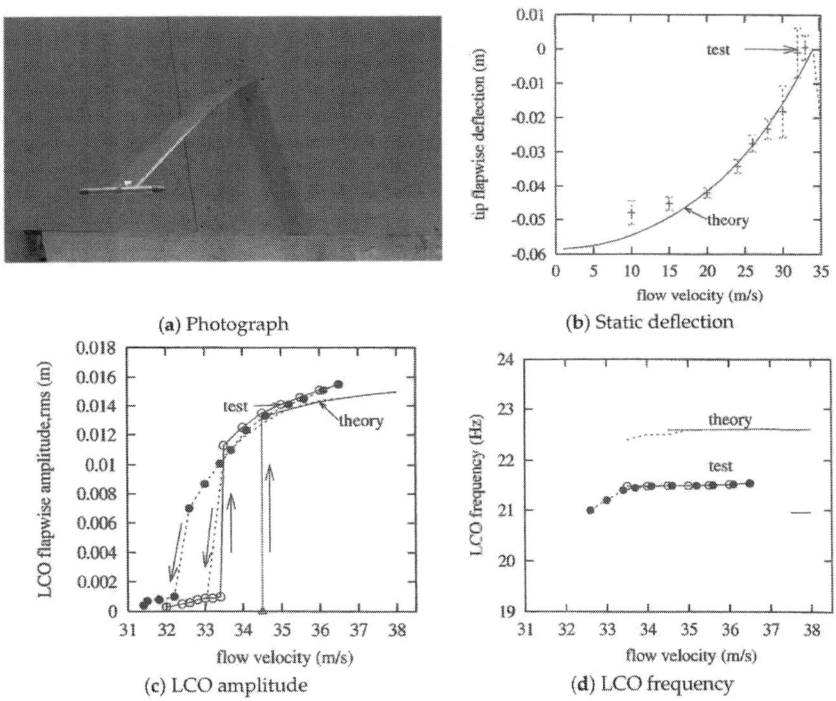

Figure 1. High altitude long endurance (HALE) experimental model and correlation analyses; (**a**) Photograph of model; (**b**) Static aeroelastic deflections at the wing tip; (**c**) Limit cycle oscillation (LCO) amplitude; (**d**) LCO frequency *vs.* flow velocity for pitch angle of attack θ0 = 1.0∘.

Axial strain gages for bending modes and a 45∘ oriented strain gage for torsional modes were glued to the root spar to measure the bending-torsional deflections of the wing. Signals from the strain gages were conditioned and amplified before their measurement through a gage conditioner and a low-pass filter. A micro-accelerometer is mounted at the mid-span of the wing. The output signals from these transducers are directly recorded on a computer with data-acquisition and analysis software, Lab-VIEW 5.1.

A Helium-Neon Laser with 0.8 mw randomly polarized and wavelength 633 (nm) is mounted on the top of the tunnel. The top of the tunnel is made of a glass plate with a thickness of 1.27 cm. A mirror, 1.27 cm in diameter, is fixed on the tip of the wing. A "mirror" deflection technique is used to determine the geometric twist angle and the vertical or flapwise bending slope at the wing tip.

All static, flutter and LCO response tests were performed in the Duke University low speed wind tunnel. The wind tunnel is a closed circuit tunnel with a test section of 0.7 × 0.53 m2 and a length of 1.52 m. The maximum air speed attainable is 89.3 m/s.

The basic parameters of the experimental wing model were obtained from standard static and vibration tests and are listed in the Table 1.

We use a "mirror" deflection technique to measure the tip static aeroelastic deflections of the wing. A point determined by a reflected light source is marked on a readout grid paper placed on the top of the wind tunnel when the wing is undeflected. The readout grid paper is calibrated in the tip flap and twist deflections before the test. The reflected light source point with wing deflection as marked on the readout paper is then determined as the flow velocity is varied.
Figure 1b shows the the theoretical and experimental results of the tip flapwise deflection for a steady angle of attack, θ0 = 1∘. The solid line indicates the theoretical results and the bar indicates the measured data. The experimental data appear have some scatter due to the turbulent aerodynamic noise, although the noise is small. A very high measurement sensitivity is obtained from the mirror technique. We use a bar in the figure to indicate the magnitude of the response uncertainty. As shown Figure 1b, the

Table 1. Experimental wing model data.

Wing Properties	
Span (L)	0.4508 m
Chord (c)	0.0508 m
Mass per unit length	0.2351 kg/m
Mom. Inertia (50% chord)	0.2056×10^{-4} kgm
Spanwise elastic axis	50% chord
Center of gravity	49% chord
Flap bending rigidity (EI_1)	0.4186 Nm^2
Chordwise bending rigidity (EI_2)	0.1844×10^2 Nm^2
Torsional rigidity (GJ)	0.9539 Nm^2
Flap structural modal damping (ξ_w)	0.02
Chordwise structural modal damping (ξ_v)	0.025
Torsional structural modal damping (ξ_ϕ)	0.031
Slender Body Properties	
Radius (R)	0.4762×10^{-2} m
Chord length (c_{SB})	0.1406 m
Mass (M)	0.0417 kg
Mom. Inertia (I_x)	0.9753×10^{-4} kgm^2
Mom. Inertia (I_y)	0.3783×10^{-5} kgm^2
Mom. Inertia (I_z)	0.9753×10^{-4} kgm^2

tip deflection increase with increasing flow velocity, but the tip deflection is alway negative until the flow velocity reaches 34.5 m/s. At that velocity the aerodynamic forces provide sufficient lift to overcome the effect of gravity. At U = 34.5 m/s, the system enters into the flutter instability range. The consequent LCO has a certain non-zero mean or temporal average and this is shown by a dashed line in Figure 1b. Also, with LCO, aerodynamic stall occurs and the lift coefficient Cl suddenly decreases. The measured data in Figure 1b are acquired before the onset of flutter and LCO. The experimental data fall near the theoretical curves in Figure 1b. In general, the agreement is good except for some points at U = 10.15 m/s. In the higher flow velocity range, the data fluctuation increases due to greater aerodynamic turbulence.

A typical LCO amplitude and LCO frequency *vs.* flow velocity for $\theta 0 = 1\circ$ and at the mid span position are shown inFigure 1c,d, respectively. The theoretical and experimental amplitudes are taken as rms average values from a 50 second sampling interval. The symbols, ∘ and •, indicate the experimental results for increasing and decreasing flow velocity, respectively. The solid and broken lines (without symbols) indicate the theoretical results for increasing and decreasing flow velocity, respectively. The symbol, Δ, indicates the linear flutter velocity which is calculated from the perturbation eigenvalue solution. For the increasing flow velocity case, the theoretical limit cycle oscillation occurs when the flow velocity is larger than the perturbation flutter velocity and the amplitude jumps from almost rest to a larger value. Once the onset of LCO occurs, the amplitude increases smoothly with increasing the flow velocity. When U> 38.6 m/s, a numerical or possibly a physical divergence is found in the theoretical model. For the case of decreasing flow velocity, as shown by the broken line, the LCO amplitude decreases, but does not exactly coincide with those for the increasing velocity case. Also, there is a jump in the LCO response at U = 33.5 m/s which is a distinctly lower velocity than that found for the increasing velocity case (*i.e.*, U = 34.5 m/s).

The present experimental and theoretical results provide new insights into nonlinear aeroelastic phenomena of LCO beyond the linear flutter velocity for high aspect ratio wings. Also, these results show the nonlinear equations and stall aerodynamic equations provided by Hodges and Dowell and ONERA, respectively, and provide useful and accurate results for HALE vehicles.

2.2. Flapping Flag and Yawed Plate Models (Nonlinear Inextensible Beam and Plate Theory)

There is a substantial literature on inextensible beam theory, but relatively little on inextensible plate theory. A notable exception is the work of J. G. Simmonds and colleagues [33,34,35,36] on inextensible plate theory. His approach is quite different from the present one and ultimately leads to a system of ordinary differential equations for the plate deformation for a one dimensional generator curve from which the entire plate deformation may then be deduced. In the present new inextensible theory of beam and plate deformation as described in [37], the formulation is based upon Hamilton's Principle in terms of a single unknown, *i.e.*, the beam or plate deformation normal to its surface. The present approach has some substantial computational advantages, but the work of Simmonds is certainly elegant and pioneering.

The interaction between a cantilevered elastic plate and a uniform axial flow is a canonical fluid-structure interaction problem. If the flow is oriented *parallel* to the clamped leading edge the system is described as a wing-like configuration. If the flow is oriented *normal* to the clamped edge then the system is referred to as flag-like. There is extensive research on the aeroelastic response of the wing-like configuration due to the similarities between the simple cantilevered beam and aircraft wings. The present study differs from the classic explorations of the swept wing due to the application of the clamped boundary condition. For a traditional swept wing configuration, the clamp is applied parallel to the flow, changing the shape of the structure for every angle (it is called the yaw angle, β, in the present paper). For the current study, the clamped edge is rotated with the structure. In this case the second classic fluid structure interaction problem for a clamped-free elastic plate will be considered, *i.e.*, the flag-like configuration.

References [12,13] focus on exploring the transition of the limit cycle oscillation (LCO) response from a small yaw angle (flapping flag response for $\beta=0$) to a large yaw angle (wing-like response, $\beta = 90^\circ$) of a cantilever plate as the angle between the clamped edge and free stream velocity changes. In particular, [37] presents a nonlinear structural model for the rotated aeroelastic system. One creates the aeroelastic model by coupling this newly developed inextensible nonlinear plate structural model with a rotated vortex lattice aerodynamic model. The theoretical predictions were validated with aeroelastic experiments. Three distinct configurations, $\beta = 0^\circ$, 45° and 90° were explored that show markedly different LCO transition behaviors.

Table 2 contains the material properties and geometry of the configuration. Table 3 shows the results of the ground vibration experiments in bending and torsion. The first column has the theoretical predictions using a ANSYS structural model with the parameters listed in Table 2. There is very good agreement between the theory (ANSYS code) and the experiment for all of the bending and torsion modes.

Table 2. Experimental plate properties.

ρ_s	E	h	L	C
2840 kg/m^3	70.6 × 10^9 N/m^2	0.381 mm	275 mm	151 mm

Table 3. Vibration test results.

Mode	ANSYS Code	Experiment	Error
	Freq (Hz)	Freq (Hz)	(Percent)
1st Bending	4.16	3.95	5.06
2nd Bending	25.84	24.98	3.3
3rd Bending	72.12	69.91	3.06
4th Bending	144.06	142.37	1.1
1st Twist	15.91	15.13	3.45
2nd Twist	45.73	49.05	7.16

The goal of the aeroelastic experiments was to quantify the LCO amplitude of the flapping flag (β = 0∘) as shown in Figure 2a. To capture the LCO amplitude, a Light Emitting Diode, LED, was placed at the tip and mid-span of the structure. Using 3D RPS video the path of the LED can be traced. Additionally the color coherence of the LED light, made it easier to use automated filtering in Matlab to track the path. To determine the amplitude of the oscillation, the code averages the snapshots from a full second of the video (30 frames) to get the average path. The script then automatically identifies the pixel of the far left and far right portion of the path and calls this the LCO amplitude. This amplitude is then

(a) $\beta = 0^0$

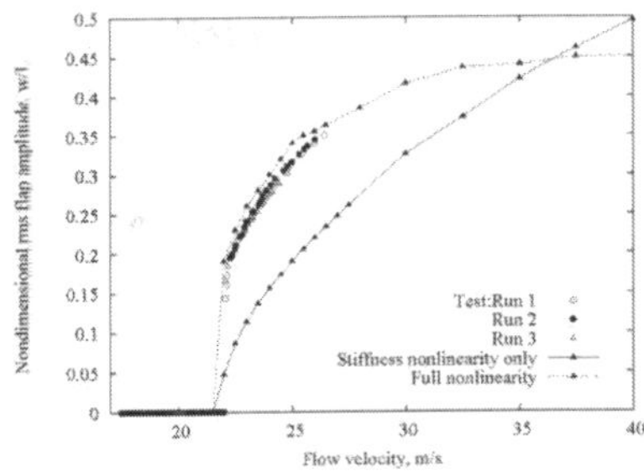

(b) **RMS amplitude**

(c) $\beta = 45^0$

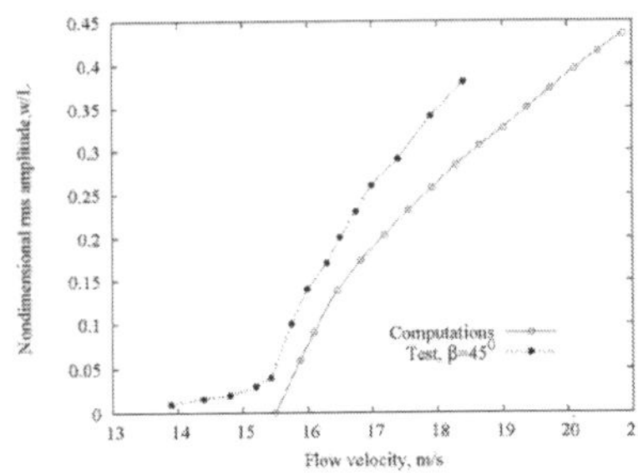

(d) **RMS amplitude**

(e) $\beta = 90^0$

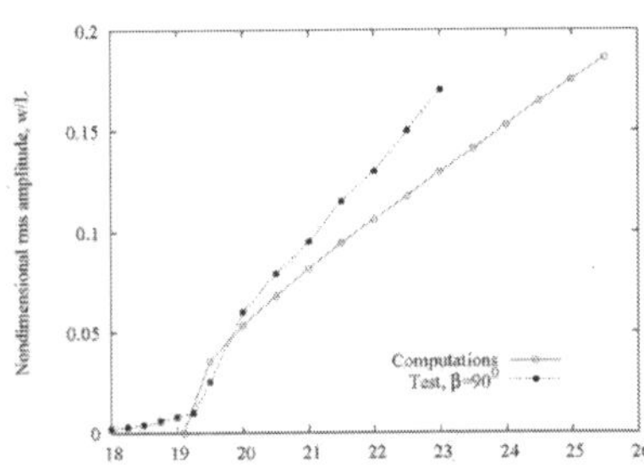

(f) **RMS amplitude**

Figure 2. Yawed plate models and correlation analyses; (**a**,**c**,**e**) for photographs and (**b**,**d**,**e**) for LCO rms amplitude versus the flow velocity for the yaw angle of β = 0∘, 45∘and β = 90∘, respectively.

synchronized with the wind tunnel velocity data to obtain the amplitude as a function of the velocity. The amplitude is converted to a real length using the camera calibration conducted before the experiment that calculates the pixels per centimeter in the plane of the LED. This method is able to provide the LCO amplitude to +/−1 centimeter. This could be improved by using a camera with a higher resolution. For the detail experimental measurement, see Reference [12].

The experimental data are presented as the measured nondimensional peak displacement from the panel tip. Figure 2b shows the nondimensional rms flap amplitude *vs.* the flow velocity. The theoretical results include two nonlinear cases: a full stiffness nonlinearity only and a full stiffness nonlinearity plus an inertia nonlinearity. The experimental results contain the LCO amplitude data for all of three different runs. The theoretical LCO responses are bursts or intermittent motion while the experimental results are primarily a single harmonic oscillation.

The experimental linear flutter speed can be estimated using an extrapolation method (this method is also used for other yaw angles), *i.e.*, an extrapolation of LCO response to zero amplitude. The experimental flutter speed is $U_f = 21.35$ m/s and the theoretical flutter speed is $U_f = 20.45$ m/s for $\beta = 0°$.

It is interesting that the theoretical results for the full nonlinearities (inertia plus stiffness) agree best with experiment for the LCO amplitude. However, when only the stiffness nonlinearity is included, the response is a single dominant frequency, as found in the experiment.

For $\beta = 45°$ and $\beta = 90°$ yawed plates, to capture the LCO amplitude, a micro-accelerometer was placed at the trailing edge of these plate structures. Even though the micro-accelerometer is only 2.6 g, we include this weight in the computational model.

Figure 2c,d show the computational and experimental correlations of the LCO displacement response rms amplitude *vs.*the flow velocity for the yaw angle of $\beta = 45°$. The correlation between the computations and experiments is reasonably good beyond the flutter speed. There are some smaller response amplitudes at lower flow velocities ($U < 15.75$ m/s) in the experiment. From the experimental curve of the LCO amplitude vs. the flow velocity, the experimental flutter speed is Uf = 15.6 m/s and the theoretical flutter speed is Uf = 16.2 m/s for $\beta = 45°$.

For $\beta = 90°$, the plate is a wing-like. Figure 2e,f show the computational and experimental correlations of the LCO displacement response rms amplitude and corresponding LCO frequency versus the flow velocity for the yawed angle of $\beta = 90°$. The correlation between the computations and experiments is fair beyond the flutter speed. There are some smaller response amplitudes at lower flow velocities ($U < 19$ m/s) in the experiment. The experimental flutter speed is Uf = 19.3 m/s and theoretical flutter speed is Uf = 18.7 m/s for $\beta = 90°$.

By comparing computations and experiments for the LCO response of the several yawed plate configurations, it is shown that the present inextensible beam or plate theory as provided by Doewll and Tang [37] produces results in good agreement with the measured rms LCO amplitude. These theories have only a single unknown variable, *i.e.*, the transverse deflection of the beam or plate structure. This is very attractive for computations.

2.3. Free-to Roll Fuselage Flutter Model (Symmetric and Anti-Symmetric Flutter/LCO Theory)

A flutter model is often made using a semi-span wind-tunnel model with the root of the semi-span wing model clamped on the tunnel side wall. The experimental data obtained from such a model is suitable for computational aeroelasticity code validation. References [38,39,40,41] are studies with typical subsonic and transonic flutter semi-span models. The boundary conditions of these experimental flutter modes are different from those of an actual aircraft, but the experimental data from the wind tunnel test may still be useful for offering the theory.

A fuselage which is free to roll is designed, but the vertical rigid body translation (plunge) and rotation (pitch) of the fuselage are constrained.The full span wing is a cropped delta wing plate for simplicity. The right and left wings are structurally symmetric. The full span wing dynamics are modeled using a linear plate wing structure theory. A three-dimensional time domain vortex lattice aerodynamic model is also used to investigate flutter of the linear aeroelastic system and results are correlated with experiment. See Reference [28] for more detail.

The experimental model consists of a right and left wing and a fuselage (slender body). For simplicity, the wing model is a cropped delta wing configuration which is constructed from a aluminum plate of thickness 0.0127 cm and has 40∘ leading edge sweep and 0∘ trailing edge sweep. The root chord is 12.7 cm and the halfspan length is 11.68 cm. The material properties of the wing are: mass density of 0.00284 kg/cm2, Young's modulus of 720,000 kg/cm2 and a Poisson's ratio of 0.3. The right and left wing model is assumed to be rigidly clamped along thefull root chord of the wing. The total rolling inertia about the fuselage center axis for the right and left wings is Jw = 3.859 × 10−5 kgm2. A photograph of the aeroelastic model in the wind tunnel is shown in Figure 3a.

The fuselage has a circular cross-section with a diameter of 2.54 cm. It includes two parts. The front part is a slender body with a parabolic forebody which can rotate about the fuselage center axis and supports the wings. The rear part is a non-rotating slender body with a parabolic aftbody which is used to support the front portion of the slender body and is connected to the wind tunnel floor by a support or sting rod. The rolling inertia about the fuselage center axis for the rotating slender body is Jf = 11.19 × 10−5 kgm2.

The wings are allowed to rotate (roll) about the center axis of the fuselage. The free or clamped mechanism for the rotation is provided by an electrical magnet brake that is used to provide rolling free or rolling clamped boundary conditions. Two strain gauges attached near the root of the two wings are used to measure the wing elastic deflections.

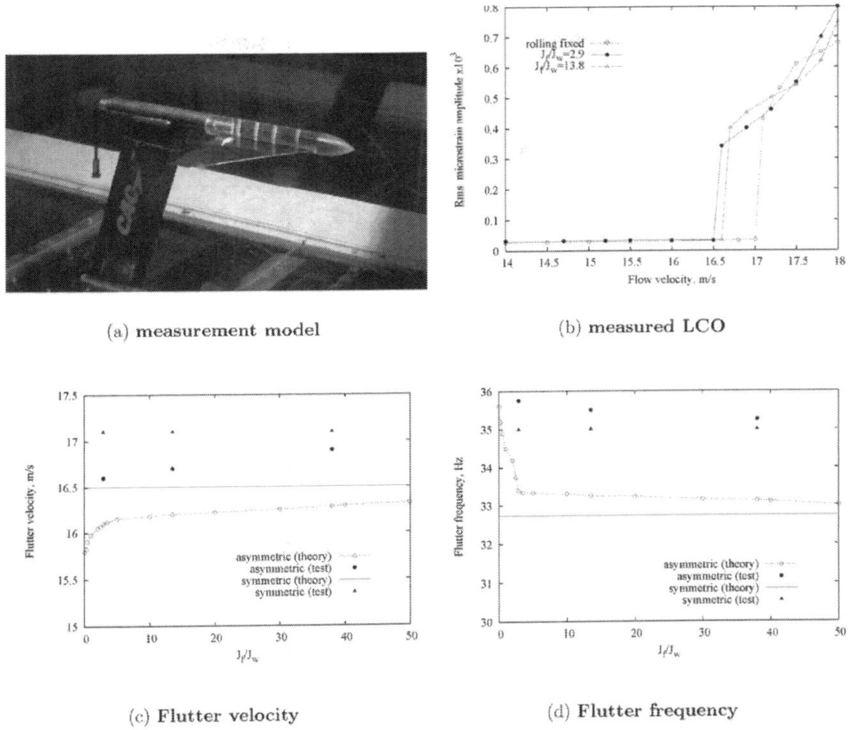

Figure 3. Free-to-roll experimental model and correlation analyses; (**a**) Photograph of the model; (**b**) Measured LCO rms amplitude *vs.* flow velocity for symmetric mode (rolling fixed) and anti-symmetric mode at Jf/Jw = 2.9 and 13.5; (**c**) and (**d**) Flutter velocity and frequency *vs.* Jf/Jw.

Because the experimental structural model has nonlinear behavior, the self-excited aeroelastic response represents a limit cycle oscillation (LCO) when the flow velocity is higher than the linear flutter speed. The linear critical flutter boundary is estimated from the LCO responses. The structural nonlinearity arises from a nonlinear strain-displacement relationship due to large structural deformations of the plate-like wings.

Figure 3b shows the measured rms LCO amplitude (microstrain) from the strain guage at the wing root *vs.* the flow velocity from U = 14 to 18 m/s for three cases. One is for the symmetric mode, *i.e.*, in the rolling fixed case. The other two cases are for the anti-symmetric mode with Jf/Jw = 2.9 and 13.5. The data are acquired when the response achieves a steady state limit cycle oscillation. The response is given by a rms value over the total sampling length. For the rolling fixed case, there are very small rms amplitudes from U = 14 to 17 m/s.It is believed that these responses are induced by the randomly fluctuating flow of the wind tunnel. At U = 17.1 m/s, the LCO rms amplitude jumps to a high amplitude response with a certain oscillating frequency. As the flow velocity continues to increase from U = 17.1 m/s, the LCO rms amplitude increases. An intersection between the extrapolated LCO amplitude curve and the flow velocity axis is defined as the linear critical flutter velocity. This estimated value is about Uf = 17.05 m/s. Similar to the symmetric mode case, for the anti-symmetric mode flutter at Jf/Jw = 2.9 and 13.5, the experimentally estimated linear critical flutter velocities are Uf = 16.6 and 16.7 m/s, respectively.

Figure 3c shows the theoretical and experimental (estimated) critical flutter boundary *vs.* Jf/Jw from Jf/Jw = 0 to 50. For the theoretical anti-symmetric mode flutter, the flutter velocity increases as Jf/Jw increases but the increase is small. For the theoretical symmetric mode flutter, the flutter boundary does not change with Jf/Jw as expected. When Jf/Jwbecomes large, the linear flutter velocity for the anti-symmetric mode approaches that for the symmetric mode. The experimental results for the anti-symmetric mode at Jf/Jw = 2.9, 13.5 and 38 are plotted in this figure as indicated by symbol of • and for the symmetric mode (rolling fixed case) is indicated by symbol of a black triangle. The maximum error is 3.5% for the flutter velocity. Figure 3d shows the corresponding flutter frequency. The maximum error is 3.5% for the flutter velocity and 7.5% for the flutter frequency.

The experimentally observed flutter mode is anti-symmetric, although theory suggests the symmetric mode flutter velocity is only modestly higher than that for the anti-symmetric modes. Correlation between theory and experiment for flutter velocity and frequency is good including the trends obtained when varying the ratio of fuselage to wing inertia.

3. EXPERIMENTAL MODELS FOR MEASURING FLUTTER/LCO RESPONSE TO EVALUATE NONLINEAR FREEPLAY THEORY

The freeplay nonlinearity usually occurs in the following components of an aircraft: (a) between the main wing and the control surface actuating mechanism; (b) between the all-movable horizontal or vertical tail and the stabilizer or control actuating mechanism. In this section, two typical experimental models have been designed and constructed to simulate these freeplay nonlinearities: (1) an airfoil section with control surface freeplay and (2) all-movable tail with stabilizer freeplay.

3.1. Airfoil Section with Control Surface Freeplay

Early theoretical studies of aeroelastic systems with structural freeplay were carried out on analog computers by linearizing the system about the nonlinearity via describing functions or harmonic balance or by time marching integration based on the state-space model. One of the compelling advantages of using harmonic balance or describing functions for aeroelastic systems is that traditional linear analytical tools, such as eigenanalysis, can be used for the determination of system stability while still allowing for the prediction of some nonlinear behavior. However, the response of a true nonlinear system may be dependent on the initial conditions, and the describing function approach does not permit a full exploration of this effect. The dependence on initial conditions and the wide variety of nonlinear behavior exhibited by systems with freeplay show the importance of incorporating this common physical nonlinearity into the theoretical model. In addition, the freeplay nonlinearity will likely have a significant effect on the response of the system to a control law designed for the nominal linear system.

In the present correlation study, a new nonlinear aerodynamic model, a finite state incompressible airloads model, developed by Peters [42] called "Duke code" and a ZAERO code based on CFD technology and nonlinear structural dynamics developed by ZONAT Trchnology, Inc. [25] have been used.

Experimental Model and Measurement System

The two-dimensional NACA 0012 rectangular wing model includes two parts: a main wing with a 19 cm chord and 52 cm span; and a flap with a 6.35 cm chord and 52 cm span, which is mounted at the trailing edge of the main wing using two pairs of micro-bearing with a pin. The main wing is constructed from an aluminum alloy circular spar beam with a diameter of 2.54 cm and a wall thickness of 0.32 cm. The beam runs through 14 pieces of NACA 0012 aluminum airfoil plate and serves as the pitch axis, located at the quarter-chord location from the leading edge. A 0.254 mm thick aluminum sheet covers the entire chord and span, providing the aerodynamic contour of the wing. In addition, an aluminum tube with 1.27 cm diameter is mounted from wing tip to wing tip at a location of 3.175 cm from the leading edge. The chordwise center of gravity can be adjusted by adding or subtracting a balance weight from this tube.

The flap is constructed in a similar manner with an aluminum alloy tube spar beam (1.27 cm diameter and 0.158 cm wall-thickness) passed through the leading edge of 14 pieces of NACA 0012 wood airfoil plate. The flap is also covered with the same type of aluminum sheet. Figure 4a shows the photograph of the experimental model including the ground vibration test.

A rotational axis comprised of the micro-bearings and pin allows the flap to have a rotational degree of freedom relative to the main wing. A steel leaf-spring is inserted tightly into a slot of the tube spar beam at one end of the flap. The free end of the leaf spring is inserted into a support block mounted on the main wing,see Figure 4b. The amount of structural stiffness that the leaf-spring provides can be adjusted by moving the support block toward or away from the rotational axis of the flap or by changing the diameter of the leaf-spring. Freeplay is incorporated by using a support block that allows the free end of the leaf-spring to move through a given range of motion before encountering resistance.

(a) **Experimental model** (b) **Photograph of flap bottom**

Figure 4. F Photograph of the experimental model (**a**) and photograph of freeplay mechanism (**b**).

The support mechanisms for the entire model are mounted outside of the wind tunnel, at the top and bottom. The support mechanism at each end is a bi-cantilever beam made of two teel leaf-springs which are 20.32 cm long, 2.86 cm wide, and 0.102 cm thick. The distance between the two cantilever beams is 15.24 cm. A support block joins the free ends of the bi-cantilevered beams on both the top and the bottom and is free to move in the plunge direction. Figure 2 shows the plunging motion of the support structure. An identical structure is located on the bottom side of the wind tunnel. The two support blocks are the only parts of the support mechanism that move with the model, and this motion is restricted to the plunge degree of freedom. The pitch axis of the main wing is mounted to the upper and lower support blocks through a pair of precision bearings which have small amount of dry friction in the ball. This design allows the model to have a plunge motion that is independent of the pitch degree of freedom. At the upper bracket, there is a spring wire inserted tightly into the pitch axis of the wing. The ends of the spring wire are simply supported on the bracket, which provides an adjustable pitch stiffness. The bracket is mounted on the upper support mechanism by three screws. The bracket allows rotation around the pitch axis when the screws are used to adjust the initial pitch angle.

The pitch angle of the main wing is measured by a rotational velocity displacement transducer, R30, which is fixed at the upper end of the pitch axis. The angular displacement transducer was calibrated and showed excellent linear response characteristic and high sensitivity. The plunge displacement is measured using another R30 which determines the motion of the upper support block. The flap rotational motion relative to the main wing is measured by a micro-RVDT which is mounted on the flap axis outside of the main wing. These three measurement signals are independent of one another.

The output from these transducers is amplified and directly recorded by data acquisition system, LabView 8.0 version in the computer. The digitized response data can be graphically displayed either on-line or off-line as a time history, phase plane plot, FFT, PSD, or Poincare map. In order to make a comparison of the theoretical and experimental data, a measurement system calibration was completed before the wing tunnel test. The dynamic calibration coefficients were determined by a ground vibration test. The system parameters for the experimental model are shown in Table 4.

From the experimental ground vibration results, a comparison of the structural natural frequencies for the numerical and experimental systems is given in Table 5.

Correlation study for zero angle of attack, $\alpha 0 = 0^\circ$

For the flutter/LCO test, a freeplay gap, $2\times\delta = 2 \times 2.12^\circ$, to be used. A nondimensional rms amplitude normalized by the freeplay gap is introduced here for LCO amplitude. The pitch and flap motions are nondimensionalized by the freeplay gap,*i.e.*, by $2\times\delta$. For the plunge motion, it is nondimensionalized by $2\times\pi\delta\, b/180^\circ$.

Table 4. Experimental plate properties.

Model Parameters:	
Chord	0.254 m
Span	0.52 m
Semi-chord (b)	0.127
Elastic axis (a) w/r/t (b)	−0.5
Hinge line (c) w/r/t (b)	0.5
Mass Parameters:	
Mass of wing	0.713 kg
Mass of aileron	0.18597 kg
Mass/length of wing-aileron	1.73 kg/m
Mass of support blocks	0.467 × 2 kg
Total mass per span, M_h	3.625 kg/m
Inertial Parameters:	
S_α (per apn)	0.0726 kg
S_β (per span)	0.00393 kg
I_α (per span)	0.0185 kgm
I_β (per span)	0.00025 kgm
Stiffness Parameters:	
K_α (per span)	46.88 kgm/s^2
K_β (per span)	2.586 kgm/s^2
K_h (per span)	2755.4 kg/ms^2
Damping Parameters:	
ζ_α (half-power)	0.0175
ζ_β (half-power)	0.032
ζ_h (half-power)	0.0033

Table 5. Model natural frequencies.

	Computational	Experimental	% Difference
ω_{α}(coupled)	8.32 Hz	8.45 Hz	1.6
ω_{β}(coupled)	17.64 Hz	17.37 Hz	1.5
ω_{h}(coupled)	4.37 Hz	4.45 Hz	1.83
U_F(m/s)	27.3	26.5	2.9
ω_F(Hz)	6.05	6.25	3.3

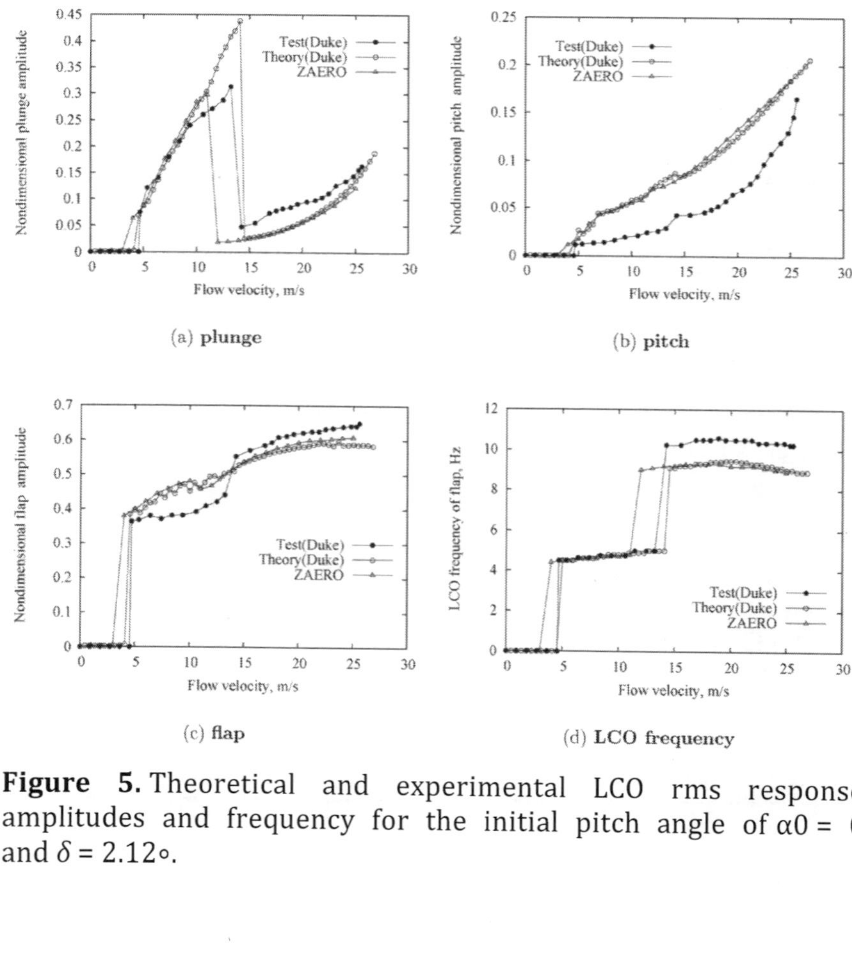

Figure 5. Theoretical and experimental LCO rms response amplitudes and frequency for the initial pitch angle of α0 = 0 and δ = 2.12∘.

When the freeplay gap is set zero, the measured flutter flow velocity is 26.5 m/s for zero initial angle of attack. Because in this flow velocity range the model motion is very large we use an external device to limit the plunge motion amplitude. The measured flutter frequency is about 6.25 Hz.

Figure 5 shows the nondimensional plunge, pitch, flap LCO amplitudes and LCO frequency *vs.* the flow velocity for the initial pitch angle, $\alpha0 = 0$. The motion may be periodic or non-periodic or chaotic. Here we use rms amplitude to present the LCO motion *vs.* the flow velocity.

As shown Figure 5, when the flow velocity, *U*, is less than 4.67 m/s no LCO occurs either for the theory or the experiment. When U is between 4.67 and 13.26 m/s for the experiment and 4.55 and 14.1 m/s for the theory, the motion is dominated by the plunge degree of freedom and has an oscillation frequency of about 4.5 to 5 Hz for both the theory and experiment. When *U* is between 13.26 and 25.6 m/s for the experiment and 14.1 and 27.4 m/s for the theory, the motion is dominated by the flap degree of freedom and has an oscillation frequency of about 10–10.5 Hz for the experiment and 9–9.5 Hz for the theory. When *U* is greater than 25.6 m/s, the experimental LCO amplitude abruptly increases and the test is stopped to protect the model (this is called the catastrophic flutter velocity). This flow velocity is close to the experimental flutter velocity (26.5 m/s) of the linear structure without freeplay. However, the theoretical catastrophic flutter velocity is 27.4 m/s and is slightly higher than the theoretical linear flutter velocity (27.3 m/s). For comparison, the computational results using CFD code, ZAERO, provided by ZONA Technology, Inc., are also presented in this figure.

Correlation study for no-zero angle of attack, $\alpha0 \neq 0$

Experimental measurement for the several initial pitch angles of attack from $\alpha0 = 2°$ to $8°$, are made to study the correlation. Here, only one case ($\alpha0 = 6°$) is shown Figure 6. Summarizing the correlation results from $\alpha0 = 2°$ to $\alpha0 = 8°$, the following observations may be made.

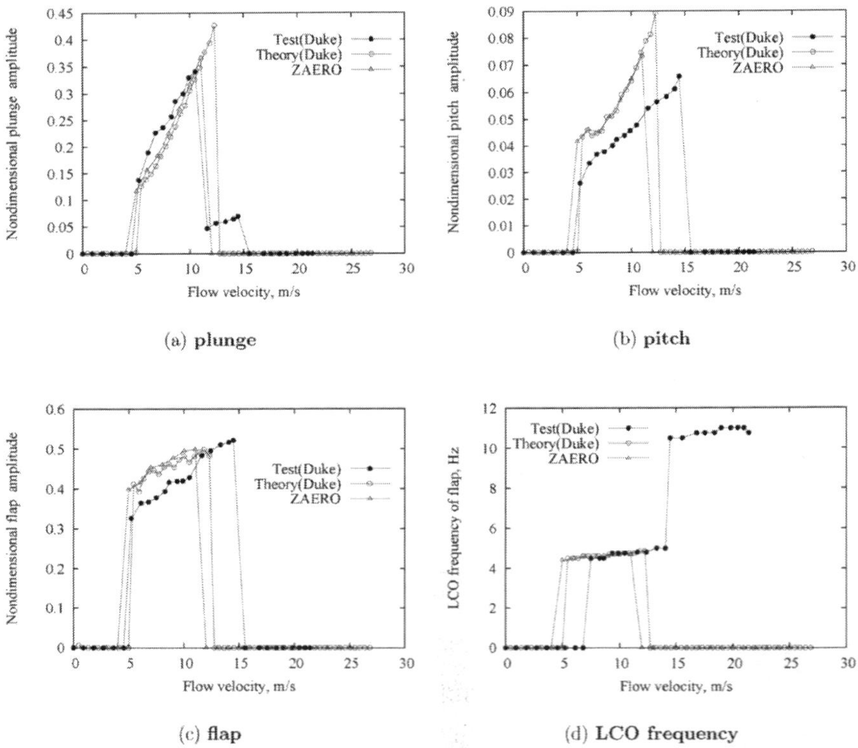

Figure 6. Theoretical and experimental LCO rms response amplitudes and frequency for the initial pitch angle of $\alpha0 = 6^{\circ}$ and $\delta = 2.12^{\circ}$.

1) The flow velocity corresponding to a change in the LCO behavior from a lower frequency oscillation to a higher frequency oscillation changes modestly in the experiments as $\alpha0$ increases.

2) The flow velocity corresponding to catastrophic flutter (very large LCO amplitude) also decreases as the initial pitch angle increases. The experimental results are 25.6 m/s for $\alpha0 = 0^{\circ}$; 25.05 m/s for $\alpha0 = 2^{\circ}$; 23.96 m/s for $\alpha0 = 4^{\circ}$; 21.48 m/s for $\alpha0 = 6^{\circ}$ and 20.43 m/s for $\alpha0 = 8^{\circ}$.

3) When $\alpha0$ equals 8°, the flow velocity range for the LCO becomes quite small and when $\alpha0$ is larger than 8° the LCO disappears both for the theory and test.

3.2. All-Movable Tail Model with Freeplay

Experimental models have been constructed by previous investigators that had free play nonlinearity in an all-movable tail in the actuating mechanism at the root, see References [43,44,45]. These early experiments were interpreted in terms of a reduction in flutter speed due to free play. However, the early works did not provide a LCO correlation analysis induced by the freeplay nonlinearity. Inspired by the early WADC (Wright Air Development Center) all-movable tail model, a similar experimental model was designed and tested. A computational code based on the linear three-dimensional time domain vortex lattice aerodynamic model and freeplay structural nonlinerarity has been developed to study the flutter and also nonlinear limit cycle oscillations (LCO) induced by a free play gap in the actuating mechanism at the root of the tail as well as the effects of the root rotation angle (nominal angle of attack) on LCO response. The tail model is mounted vertically in the wind tunnel. The effect of the gravity load is not considered in the computations and experiment discussed here although a later study included the effects of gravity by testing the model in a horizontal mount.

3.2.1. Experimental Model and Measurement System

A uniformly tapered wing from root to tip was designed to simulate a conventional unswept horizontal stalbilizer. The schematic of the all-movable horizontal tail model with free play gap in the root rotation mechanism is shown in Figure 7a. The experimental model includes two parts: a tapered wing and an actuating mechanism with torsional free play gap at the root. The tail wing is untwisted and flexible in the flap, lag and torsional directions. The root chord is 24.7 cm and tip chord is 13.2 cm and the span is 39.7 cm. The sweep angle of the lead edge is 5.7 ∘.

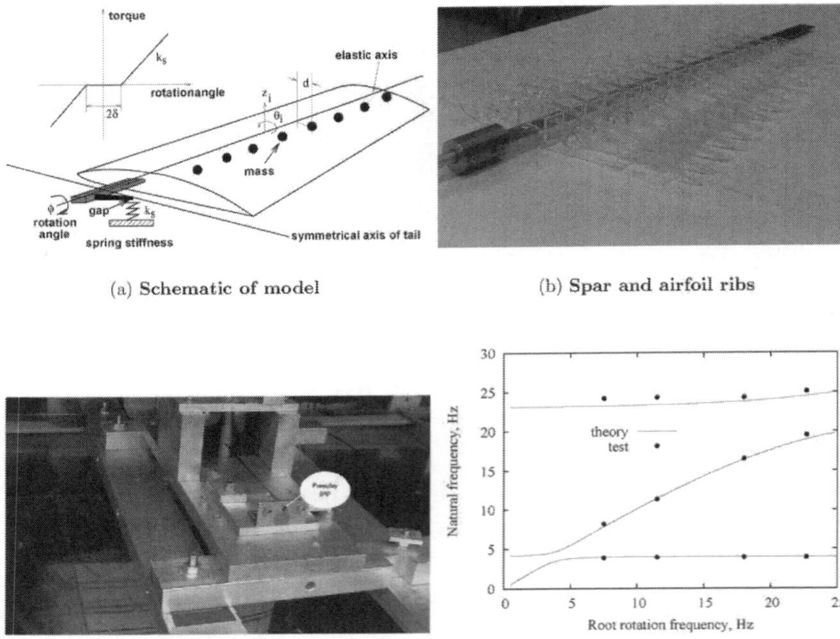

(a) Schematic of model

(b) Spar and airfoil ribs

(c) Freeplay mechanism

(d) Natural frequency

Figure 7. (**a**) Schematic of the all-movable tail model with free play in the root rotation mechanism; (**b**) and (**c**) Photographs of the model configuration and free play at root support mechanisms; (**d**) Natural frequency *vs.* uncoupled rotation frequency, $\omega\phi$ at the root.

The wing is constructed from a precision ground flat steel spar with mass uniformly distributed along the wing span. The spar is 39.7 cm in length, 1.905 cm in width and 0.127 cm in thickness. It is inserted tightly into the wing root mechanism. The spar elastic axis is located at 33 percent of the local chord aft of the leading edge which is consistent with the location for the axis of rotation of the tail, see Figure 7b.

There are 14 pieces of 2.6 cm wide sections uniformly attached to the spar along the span. Each section consists of a pair of ribs with NACA 0012 airfoil plate made of plastic material with 0.15 cm thickness. Each airfoil plate is fabricated using a Laser Printer and CAD software. Each airfoil section is constructed from two plastic airfoil plates and balsa wood. The wing is assembled by inserting the steel spar through a small slot cut through each of the sections. An additional 0.2 cm thick sheet of balsa wood is placed between each section and the wing is permanently bonded. The balsa wood placed between the airfoil sections provides some additional mass and stiffness in the bending and in torsion.

The support mechanism for the entire tail model are mounted outside of the wind tunnel at the top. The wing spar axis is mounted to support blocks through a pair of precision bearings which have a small amount of dry friction in the ball, but this dry friction is not accounted in the computational code. There is a spring wire inserted tightly into a rotational axis between two bearings. The free end of the spring wire is simply supported on a bracket of the support block. The amount of rotational stiffness that the spring wire provides can be adjusted by moving the support block toward or away from the rotational axis of the wing. Each free play gap is incorporated by using a support block with a different size hole that allows the free end of the spring wire to move through a given range of motion before encountering resistance. A photograph of the spring wire assembly for the rotation stiffness (with free play) is shown in Figure 7c. The bracket is mounted on the support mechanism by screws. The bracket allows rotation around the pitch axis when the screws are used to adjust the nominal pitch angle.

The root pitch angle of the wing is measured by a rotational velocity displacement transducer, R30, which is fixed at free end of the support block. The angular displacement transducer was calibrated and showed excellent linear response characteristic and high sensitivity. A 45° oriented strain gage for torsional modes were glued to the root spar to measure the torsional elastic deflections of the tail. Signals from the strain gages were conditioned and amplified before their measurement through a gage conditioner and a low-pass filter. A micro-accelerometer is mounted at the tip-span of the tail. The output signals from these transducers are directly recorded on a computer with data-acquisition and analysis software, Lab-VIEW 7.1.

A ground vibration test of the model was used to determine the natural frequencies and the modal damping. Figure 7d shows the theoretical and experimental natural frequencies at zero airspeed *vs.* the uncoupled rotation frequency, $\omega\phi$. The first mode is dominated by the first bending and the second is dominated by the first torsion. The third mode is dominated by the second bending. When the tail wing is cantilevered, $\omega\phi=\infty$, the first two theoretical bending frequencies are 3.99, 23.1 Hz and the first torsion frequency is 30.2 Hz. The experimental results are 4.0, 25.1 for bending and 27.75 Hz for torsion. In this case the agreement between the theory and experiment is good. Also it is found that the frequencies are close to the corresponding $\omega\phi=\infty$ values (as shown in Figure 7d) over the higher $\omega\phi$ range. This means the coupling between the bending and torsion modes is weak.

3.2.2. LCO Correlation Study for $\alpha 0 = 0°$

Four typical free play gaps are selected. These are $\delta = 0.1°$, $0.25°$, $0.5°$ and $1°$. The military specification (MIL-SPEC) confines the free play limit of all-moveable control surfaces to be less than $0.034°$ (peak to peak value) to achieve the goal of no free play induced LCO during normal operation of aircraft. In the present study, the maximum free play gap is limited to $\pm 1°$. The uncoupled rotation frequency, $\omega\phi$, is chosen to be 7.5 Hz, corresponding to root rotation stiffness, ks = 2.895 Nm. Considering the effects of flight loads on tail free play induced LCO, the initial angles of attack vary from $\alpha 0 = 0°$ to $6°$. The purpose of these computations is to determine the nominal angles of attack for which when the free play induced LCO disappears.

Depending on the initial conditions and physical parameters, the response may be either LCOs or chaotic oscillations, *i.e.*, the oscillation is generally around one of two equilibrium positions ($\pm\delta$), and the particular equilibrium position depends upon the initial conditions. Thus the oscillating displacement is essentially random with respect to time. In this study, we used a velocity response for the bending motion and a torsional rate for the torsional motion to present the LCO response behavior. A rms amplitude for the bending displacement velocity at the tail tip and a rms amplitude for the rotation angle at the root are calculated.

Figure 8a,b show the rms LCO amplitude of the bending velocity and at the tail tip (a) and the rms LCO amplitude of the root rotational angle motion (b) *vs.* the flow velocity for the initial disturbance angle, +δ at the root. The flow velocity range is from 0 to 22 m/s. Note that the linear flutter speed without free play is 23 m/s. In these figures results are included for four free play gaps, $\delta = 0.1°$, 0.25°, 0.5° and 1°. From Figure 8, the following observations may be made.

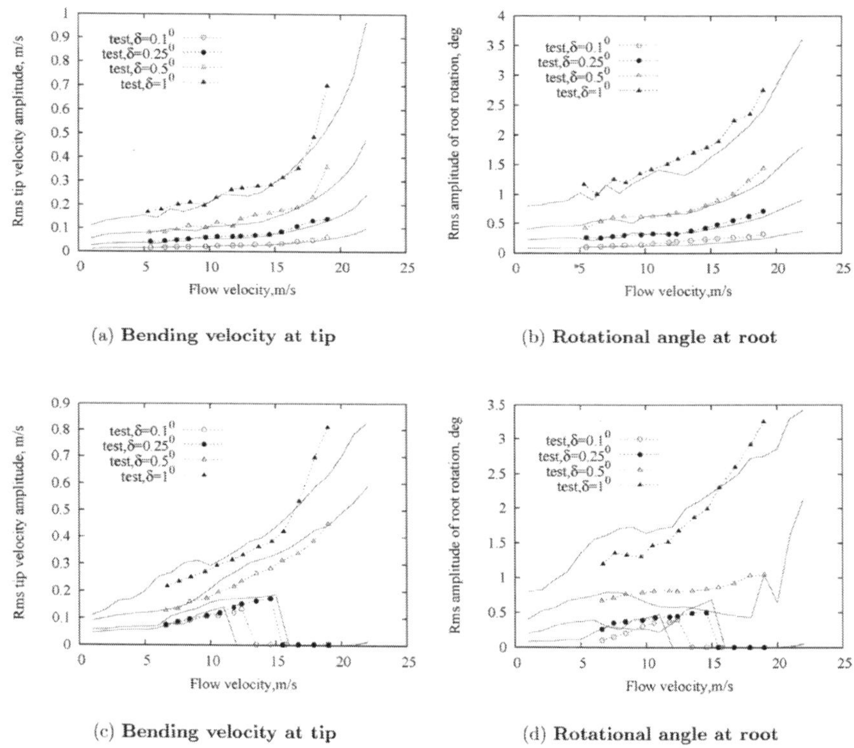

Figure 8. Correlation analyses for the all-movable model; (**a**) Tip bending velocity and (**b**) Root rotational angle *vs.* the flow velocity for α0 = 0° and (**c**,**d**) for α0 = 2°.

1) The theoretical LCO occurs over essentially the whole flow velocity range for which when U>0. There is an oscillating divergence from the LCO motion when the flow velocity is greater than the linear flutter speed, *i.e.*, U≥Uf. The experimental LCO occurs when flow velocity is larger than about 5 m/s. Below 5 m/s there is no measured LCO due to a dry friction torque in the root rotation mechanism. Without

this dry friction torque, LCO is predicted to occur at even lower flow velocities.

2) The rms amplitude for both the bending and torsional response increases as the flow velocity increases.

3) The rms amplitude for the root rotational angle response oscillates around the gap region. The amplitude varies with the flow velocity which depends upon the initial conditions.

4) As expected the rms amplitude for both the bending and torsional response increases as the free play gap increases.

3.2.3. LCO Correlation Study for $\alpha 0 \neq 0^\circ$

Five typical nominal angles of attack, $\alpha 0$ = 1°, 2°, 4°, 6° and 8° and four free play gap values as described before. Only one case is presented here. Figure 8c,d show the rms LCO amplitude for a nominal angle of attack, $\alpha 0$ = 2°. For two free play gaps, δ = 0.1° and δ = 0.25° LCO disappears above a certain flow velocity. The corresponding flow velocities for which LCO disappears, Uno–LCO, are 12 m/s for the theory and 13.5 for the test in the case of δ = 0.1° and 16 m/s for the theory and 14.6 m/s for the test in the case of δ = 0.25° respectively. When δ > 0.25°, LCO response occurs over the whole flow velocity range Uf>U>0.

Summarizing the theoretical and experimental results for $\alpha 0$ = 1° to 8°, a disappearance of the LCO boundary is listed inTable 6. Note that the values in the parenthesis show the results from the tests. Also as the nominal angles of attack increases from 1° to 8°, the following observations may be made.

Table 6. Flow Velocity at which LCO Disappeared, Uno–LCO m/s.

α_0/δ	0.1°	0.25°	0.5°	1.0°
1°	17/(15.7) m/s	None	None	None
2°	12/(13.5) m/s	16/(14.6) m/s	None	None
4°	8/(0) m/s	13/(11) m/s	16/(14.1)	None
6°	0/(0) m/s	10/(0) m/s	15/(13)	17/(14.7)
8°	0/(0) m/s	0/(0) m/s	13.5/(0)	15.5/(12)

1) The flow velocity for LCO disappearing increases when the angle of attack decreases for a given free play gap.
2) The free play boundary for LCO disappearing increases when the angle of attack increases for a given flow velocity.
3) As expected the LCO rms amplitude for both the bending and torsional responses increases with increased free play gaps. These results are similar to those for $\alpha 0 = 0°$.
4) Note that LCO also disappears for flow velocities less than about 6.5 m/s which is attributed again to the presence of friction in the root support mechanism.

The quantitative agreement between the theory and experiment is reasonably good for zero and non-zero angles of attack. The fair to good quantitative agreement between theory and experiment verifies that the present method has reasonable accuracy and good computational efficiency for the flutter/LCO analysis.

4. EXPERIMENTAL MODELS FOR MEASURING AERODYNAMIC RESPONSE PHENOMENON IN BUFFETING FLOW

In earlier computational work by Raveh and Dowell, References [46,47], large shock oscillations (buffet oscillations) are observed for a certain combination of Mach number and steady mean angle of attack even in the absence of structural motion. Also in lower Reynolds number flow, bluff bodies in cross flow [48], and a flat plate at high angles of attack [49] are observed to have the nonlinear oscillations induced by the buffeting flow. In the buffeting flow there is a very interesting phenomenon called "frequency lock-in". Under certain flow conditions, the shedding frequency can lock into the natural frequency of the structure. In this Lock-in region, the structural response amplitude can reach a limit cycle oscillation that may cause high cycle fatigue and eventually failure.

Two experimental models were constructed to observe the buffeting flow oscillation. One is to find lock-in region by measuring the pressure variation on the oscillating airfoil surface at high angles of attack. Another is to construct a partial-span control surface (flap) attached to near the mid-span of a NACA0012 airfoil model with and without freeplay gaps at high angle of attack and to measure the flap response induced by the buffeting flow.

4.1. An Oscillating Airfoil Section Model in Buffeting Flow

4.1.1. Experimental Model and Measurement System

In order to observe the dynamic and static pressure distribution and measure the aerodynamic loading due to a high nominal angle of attack and a small pitching oscillation through wind tunnel tests, a typical section wing model was constructed. Design of the typical section took into account the fact that it must accommodate a root pitching oscillation, mount 20 micro-pressure sensors, PCB model 103B01, for dynamic pressure measurement and 20 orifices for the static pressure taps at the mid-span and also be mounted in the Duke University wind tunnel using an existing pitch oscillating testing rig.

The profile of the typical section is a NACA 0012 symmetrical airfoil, having a chord length of 0.2554 m, which nicely accommodates 20 dynamic pressure sensors and 20 static pressure taps. The span, l = 0.52 m of the typical section was dictated by the height of the wind tunnel, as the typical section is to be mounted vertically. Eight pieces of NACA0012 airfoil ribs made of aluminum material with 0.64 cm thickness are used for high chordwise stiffness and for adequate housing for bearings supporting the small strip hinge. In addition, a mid-span aluminum rib with 1.27 cm thickness provides support for 20 pressure sensors, see Figure 9a. The airfoil ribs are placed on three spars that, in addition to being structural elements, have specific functional roles.The spar placed at the quarter chord is an aluminum tube and serves to mount the model on the testing rig that is connected to a pitch oscillating testing rig to create an airfoil pitch oscillation.

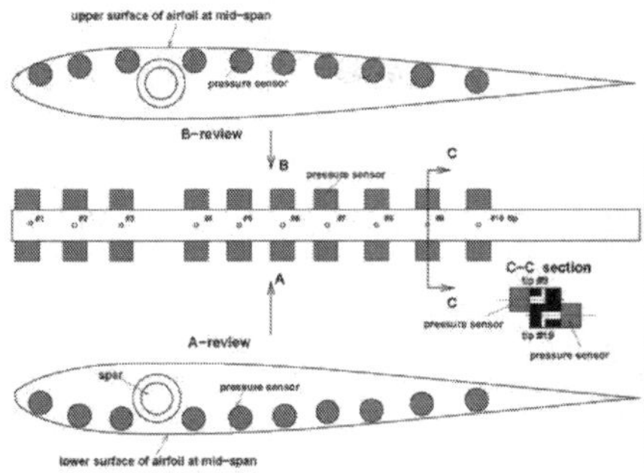

(a) **Schematic of pressure measurement**

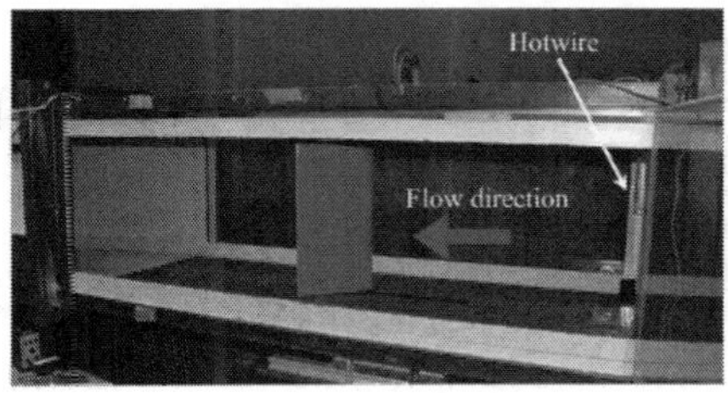

(b) **Photograph of model**

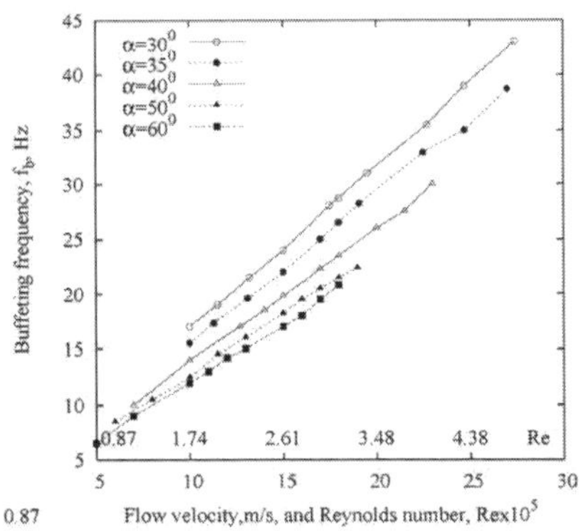

(c) **buffeting frequency**

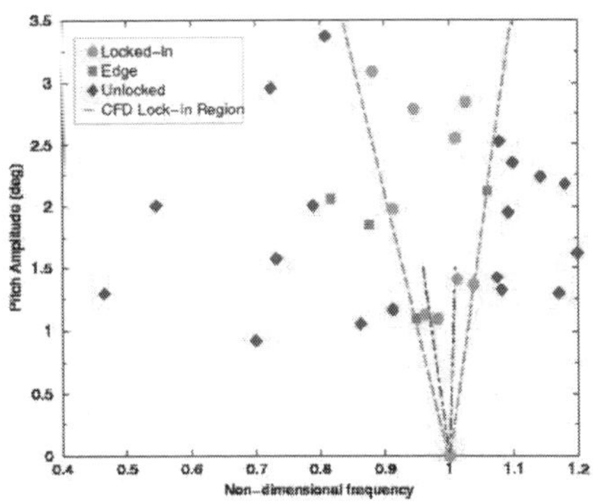

(d) **lock-in region**

Figure 9. Oscillating airfoil model in the buffeting flow and correlation analyses; (**a**) Schematic of pressure measurement; (**b**) Experimental model in wind tunnel at high angles of attack; (**c**) Buffet frequency varies flow velocity and Reynolds number, Re, for "onset" boundary of the buffet flow for nominal angles of attack, α = 30∘ to 60∘, (**d**) Experimental and theoretical lock-in region for α = 40∘ and the Reynolds number, Re= 100,000 (U = 5.8 m/s).

The mid-span airfoil rib is specially made to measure the dynamic pressure distribution at mid-span and provide support for the 20 micro-pressure sensors. There are 20 orifices of 1.16 mm diameter for the pressure taps which are symmetrically distributed over the upper and lower surfaces, respectively. The orifice positions in terms of percentage chord from the leading edge are 0.06, 0.12,

0.18, 0.3, 0.42, 0.49, 0.54, 0.61 and 0.74. The pressure sensor is bolted on the mid-span airfoil rib and installed with an adhesive mounting ring to prevent a pressure leak between the pressure surface of the sensor and the airfoil orifice. The dimensions of the pressure sensor are 9.5 mm diameter, 7 mm height. The weight is 3.26 g. The signal/noise is 60 db, *i.e.*, the uncertainty is 0.1%. Th dynamic range is 77 db and the frequency range is 5 Hz to 13 kHz. Details of the dynamic pressure measurement device are shown in Figure 9a. All the structural elements are held together by epoxy resin and mechanical elements (bolts, pins, *etc.*). They are covered with 0.25 mm in thick aluminum foil (called skin).

The root airfoil pitch excitation was performed by a pitch oscillatory shake table which is mounted to a very heavy support frame that is attached to the ground. The shake table is driven using a DC servo motor through a cam.The driving frequency is controlled by a D/A NB-MIO-16 system. A nearly pure single harmonic excitation is provided. A photograph of the wind tunnel model and the pressure measurement system in the model is shown in Figure 9b.

For the dynamic pressure measurement, the output of each pressure sensor and pitch oscillating displacement was directly recorded on a PC computer through signal conditioners, PCB models481A02, 442C04 and a data acquisition package which consisted of a 20-channel analog to digital (A/D) plug-in interface board, a BNC termination box,and data acquisition and analysis software, LabVIEW version 9.0.

4.1.2. Correlation Analyses for Frequency Lock-in Region

In this section, for comparison with the experimental results, the computational fluid dynamic analyses of NACA0012 are conducted using a nonlinear, frequency domain, harmonic balance (HB) code [50]. The viscous, Reynolds-averaged Navier-Stokes equations are solved for a 2D slice using a Spalart-Allmaras turbulence model. The O-grid contains 385 points in the circumferential direction, 97 points in the radial direction and extent approximately 20 diameters from the airfoil.

Aerodynamic response at buffet flow conditions with no pitch excitation were measured for different flow velocities from U = 5 to 27.5 m/s and several nominal angles of attack, α = 30° to 60°. The results are shown in Figure 9c in a plot of buffeting frequency *vs.* flow velocity and corresponding Reynolds number for the "onset" boundary of the buffet flow.

For aerodynamic response to pitch excition in the buffeting flow, a pitch sinusoidal excitation at the airfoil root is added. The pitch excitation amplitudes, θ0, are considered in the present experiment from θ0 = 0.8° to 3.4°. The excitation frequency, fe (ωe=2πfe) varies from 5 to 12.5 Hz for the nominal angle of attack, α = 40° and the flow velocity, U = 5.8 m/s and Reynolds number Re = 100,000.

Figure 9d shows the experimental lock-in region found by enforcing the airfoil oscillations at a fixed amplitude and frequency. The non-dimensionalized frequency is defined as fe/fb. The angle of attack is 40° and the Reynolds number is 100,000 (U = 5.8 m/s). Figure 9d and other comparable data may be found in Besem *et al.* [51]. The diamond symbols represent conditions where the shedding and enforced frequencies are unlocked, the circle symbols show lockin conditions, and the square symbols have a chaotic behavior that can be attributed to the edge of the lock-in region. The "V" shape is similar to the numerical results conducted by HB code [50], though the lock-in region is narrower numerically, which can be attributed to the chaotic behavior on the edges of the lock-in region. In the experiments, it is difficult to separate the frequencies due to the motion and due to the shedding because of the noise in the measurements. However, the computational code solution will only converge if the two frequencies are perfectly locked-in. This explains why the edges of the lock-in region are difficult to compare with the experiments.

From Figure 9c,d the following observations may be made.

1) The buffeting frequency is almost directly proportional to the flow velocity or Reynolds number for any nominal angle of attack.

2) The buffeting frequency (the starting points of the curves) depends on a combination of nominal angle of attack and flow velocity or Reynolds number.

3) As the nominal angle of attack increases the buffeting frequency decreases for a certain flow velocity or Reynolds number.

4.2. An Airfoil with and without Freeplay Control Surface in Buffeting Flow

4.2.1. Experimental Model and Measurement System

The airfoil experimental model and the pressure measurement system are the same as described above. This model is mounted vertically to a very heavy support frame that allows one to adjust the nominal angle of attck of this model. The rotation center is at the one quarter chord of the airfoil. A photograph of the experimental model is shown in Figure 10a.

Based on the above model, a partial span control surface (flap) with 5.08 cm chord and 5.08 cm span is mounted on the wing model at span location, y = 28 cm and chord location of x = 20 cm. The flap pitch rotation axis is placed at x = 20 cm. The rotation axis is comprised of two micro-bearings and allows the flap to have a rotational degree of freedom relative to the main wing. A steel leaf-spring is inserted tightly into a slot of the rotation axis of the flap. The free end of the leaf-spring is inserted into a support block mounted on the main wing. The amount of structural stiffness that the leaf-spring provides can be adjusted by moving the support block toward or away from the rotational axis. Freeplay is incorporated by using a support block that allows the free end of the leaf-spring to move through a given range of motion before encountering resistance. A schematic of the leaf spring assembly for the partial control surface (with freeplay) is shown in Figure 10b.

Two strain gauges are glued to both sides near the fixed end of the leaf-spring and are used to measure the flap pitch angle. The pitch angle calibration has been made from the strain gages. There is almost a linear relationship between the output voltage and the input pitch angle within 15 degrees. The average sensitive coefficient is 1.115 degree/voltage. For the dynamic pressure and flap pitch angle measurements , the output of each pressure sensor

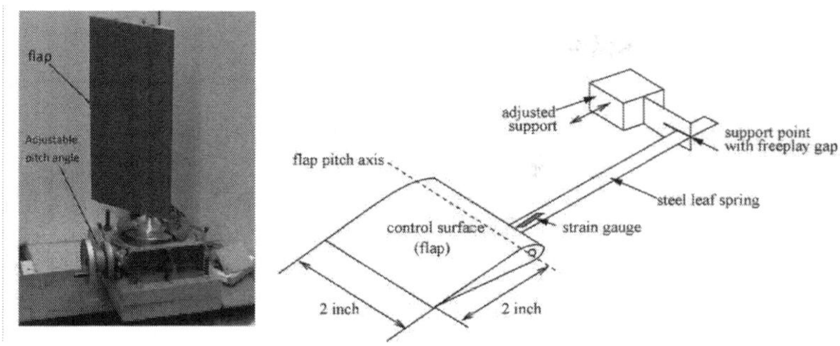

(a) **Photograph of model** (b) **Schematic of flap**

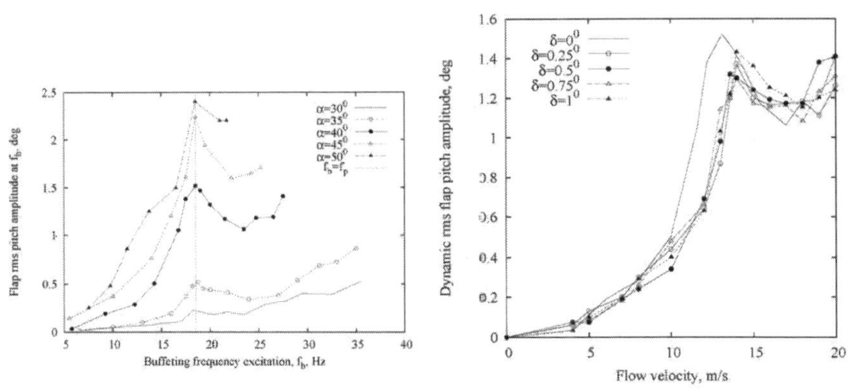

(c) **Rms amplitude vs.** f_b, $\delta = 0$ (d) **Rms amplitude vs. U**, $\delta \neq 0$

Figure 10. Experimental Freeplay model and correlation analyses in buffet flow; (**a**) Photograph of model; (**b**) Schematic of a portion of main wing-partial span surface assembly; (**c**) Dynamic rms flap pitch amplitude *vs.* buffet frequency for α = 30∘ to 50∘without freeplay; (**d**) Dynamic rms flap pitch amplitude *vs.* flow velocity for α = 40∘and freeplay gap, δ = 0.25∘ to 1.0∘.

and pitch angle were directly recorded on a PC computer through signal conditioners and a data acquisition package which consisted of a 20-channel analog to digital (A/D) plug-in interface board, a BNC termination box, and data acquisition and analysis software, LabVIEW version 9.0.

Because of unavoidable freestream and boundary layer turbulence irregularities, the free stream turbulence levels are about 1% of the nominal velocity. In order to remove this randomness from the pressure data, an ensemble averaging procedure over a total of 20 records was carried out. The response is given by a rms value over the total sampling length for the experimental data. For the FFT analysis of these experimental data, an ensemble averaged FFT analysis has been used. Twenty (20) test time samples are taken to do the FFT analysis and an averaged FFT amplitude and frequency are determined.

4.2.2. Measured Aeroelastic Response of the Flap Induced by Buffeting Flow

The experimental flap pitch stiffness is $k_p = 0.0569$ Nm/rad and the flap pitch moment of inertia is $J_p = 0.42155 \times 10^{-5}$ kgm^2 and the corresponding flap pitch natural frequency is $f_p = 18.25$ Hz. All aeroelastic flap response tests were performed for nominal angles of attack from $\alpha_0 = 30^\circ$ to 50° and the flow velocity varies from 5 m/s to 26 m/s. Four freeplay gap ranges are considered in the wind tunnel tests, $\delta = \pm 0.25^\circ, \pm 0.5^\circ, \pm 0.75^\circ$ and $\pm 1.0^\circ$.

Figure 10c shows flap rms pitch amplitude *vs.* the buffeting frequency, for the nominal angles of attack $\alpha_0 = 30^\circ$ to 50°without freeplay gap. There is a local peak amplitude for each nominal angle of attack. Corresponding to these peak amplitudes, the flap is excited by a larger aerodynamic load with a certain excitation frequency, *i.e.*, the buffet frequency, f_b.(Recall that the buffet frequency is given in Figure 9c as a function of flow velocity for the various angles of attack.) At the peak response values, the buffeting frequencies are close to the flap natural frequency, f_p. We use a dotted line to connect these peak points in this figure. The rms dynamic flap responses are created by two different types of flow separation: one is due to a randomly fluctuating flow as shown for $\alpha_0 < 28^\circ$ and the other is due to a von Karman street vortex flow that has a dominant frequency when $\alpha_0 > 30^\circ$. The fluctuating flow has a white noise characteristic and the noise strength increases as the flow velocity increases. This noise excitation always leads to an increase in rms flap amplitude with increased flow velocity, but the von Karman street vortex flow or buffeting flow significantly increases the rms flap amplitude when the buffeting frequency is close to the flap natural frequency.

Generally, the rms flap response amplitude increases as the nominal angle of attack increases. For example, when $\alpha 0 = 30°$, the buffet flow strength, *i.e.*, the peak value at the buffeting frequency, is relatively weak. Indeed, for the range of $28° < \alpha 0 < 30°$ it is difficult to identify the dominant buffeting frequency from the measured flap response due to the weak buffeting flow. When $\alpha 0 > 40°$ the buffet flow strength is stronger and the effect on the flap response is significant.

The dynamic rms amplitude of the flap pitch angle *vs.* the flow velocity for four freeplay gaps at nominal angle of attack, $\alpha 0 = 40°$ is shown in Figure 10d. The dynamic rms amplitude does not increase monotonically with an increased flow velocity. There is a local peak amplitude for each freeplay gap. Corresponding to these peak amplitudes, the flap is excited by a larger aerodynamic load with a buffeting flow frequency, fb and these buffeting frequencies are close to the flap natural frequency, fp. Corresponding to the peak amplitude, the flow velocity, Up−p = 13.1 m/s and frequency fb = 18.5 Hz for no freeplay gap. The Up−p values for different freeplay gaps are also very close to *U* = 13.1 m/s. The corresponding fp−p values for different freeplay gaps are very close to the buffeting frequency, fb. It is interesting to note that the peak rms amplitude for different freeplay gaps are also close to the value for no freeplay case. This means the effect of the freeplay gap on the dynamic response of the flap motion is small.

Due to the larger static preload of the aerodynamic forces, the flap pitch motion is always near to one side of the freeplay boundary. The flow buffeting frequency *vs.* the flow velocity can be determined from measured flap motion. The results show that the flow buffeting frequency is almost independent of the freeplay gaps, as expected.

The buffet flow occurs at large angles of attack and leads to a larger preload due to the aerodynamic forces and thus a larger static pitch equilibrium position. In this case no limit cycle oscillations (LCO) of the flap pitch motion induced by the freeplay gap were found. The flap pitch motions are around the upper boundary of the freeplay gap only. Buffeting flow excitation leads to a large dynamic response when the buffeting frequency is close to the flap's natural frequency.

5. CONCLUSIONS

This overview of wind tunnel model design and testing conducted at Duke University over the last twenty years has demonstrated that a standard low speed wind tunnel may be used as a test bed for a wide range of models to investigate linear and nonlinear phenomena in aeroelasticity. The data acquired have been used to validate and improve theoretical/computational models that in turn have provided a basis for high speed flow prediction of comparable phenomena. Flutter, gust response and limit cycle oscillations have been among the fundamental physical phenomena that have been studied for models with a range of nonlinear elements in the structures (including freeplay and nonlinear geometric, *i.e.*, strain-displacement relations) or aerodynamic flows (including flow separation over bluff bodies and airfoils at high angles of attack). Correlation between theory and experiment has been encouraging and has continually improved as the state of the art has progressed.

ACKNOWLEDGMENTS

The authors would like to thank all those colleagues who have contributed to these studies over the years including the following: Peter Attar, Fanny Besem, Ping Chih Chen, Mark Connor, Chuck Denegri, Henri Gavin, Chad Gibbs, Adam Grasch, James Henry, Denis Kholodar, Bob Kielb, Dong-Hwan Lee, Daniella Raveh, Anosh Sethna, Meredith Spiker, Jeff Thomas, Steve Trickey, Lawrie Virgin, Ivan Wang, Hiroshi Yamamoto, Minghui Zhao.

AUTHOR CONTRIBUTIONS

Deman Tang and Earl H. Dowell contributed to the aeroelastic model design, wind tunnel testing for correlation with new theory and preparation of this manuscript.

CONFLICTS OF INTEREST

The authors declare no conflict of interest.

REFERENCES

1. Tang, D.M.; Dowell, E.H. Experimental and Theoretical Study on Aeroelastic Response of High-Aspect Ratio Wings.*AIAA J.* **2001**, *39*, 1430–1441.
2. Tang, D.M.; Dowell, E.H. Limit Cycle Hysteresis Response for a High-Aspect Ratio Wing Model. *J. Aircr.* **2002**, *39*, 885–888.
3. Tang, D.M.; Dowell, E.H. Experimental and Theoretical Study of Gust Response for a High-Aspect Ratio Wing. *AIAA J.* **2002**, *40*, 419–429.
4. Tang, D.M.; Grasch, A.; Dowell, E.H. Gust Response for Flexibly Suspended High-Aspect Ratio Wing. *AIAA J.* **2010**,*48*, 2430–2444.
5. Tang, D.M.; Dowell, E.H. Flutter/LCO Suppression of High Aspect Ratio Wings. *J. R. Aeronaut. Soc.* **2009**, *113*, 409–416.
6. Tang, D.M.; Henry, J.K.; Dowell, E.H. Limit Cycle Oscillations of Delta Wing Models in Low Subsonic Flow. *AIAA J.***1999**, *37*, 1355–1362.
7. Tang, D.M.; Dowell, E.H. Effects of Angle of Attack on Nonlinear Flutter of a Delta Wing. *AIAA J.* **2001**, *39*, 15–21.
8. Tang, D.M.; Attar, P. J.; Dowell, E.H. Flutter/LCO Analysis and Experiment for a Wing-Store Model. *AIAA J.* **2006**, *44*, 1662–1675.
9. Tang, D.M.; Dowell, E.H. Flutter and Limit Cycle Oscillations for a Wing-Store Model with Freeplay. *J. Aircr.* **2006**, *43*, 487–503.
10. Attar, P.J.; Dowell, E.H.; Tang, D.M. Modeling the LCO of a Delta Wing Model with an External Store Using a High Fidelity Structural Model. In Proceedings of the AIAA SDM Conference, Austin, Texas, 18–21 April 2005.

11. Tang, D.M.; Yamamoto, H.; Dowell, E.H. Experimental and Theoretical Study on Limit Cycle Oscillations of Two-Dimensional Panels in Axial Flow. *J. Fluids Struct.* **2003**, *17*, 225–242.

12. Gibbs, C.; Sethna, A.; Wang, I.; Tang, D.M.; Dowell, E.H. Aeroelastic Stability of a Cantilevered Plate in Yawed Subsonic Flow. *J. Fluids Struct.* **2014**, *49*, 450–462.

13. Tang, D.M.; Gibbs, C.; Dowell, E.H. Nonlinear Aeroelastic Analysis with Inextensible Plate Theory Including Correlation with Experiment. *AIAA J.* **2015**, *53*, 1299–1308.

14. Attar, P.J.; Dowell, E.H.; Tang, D.M. A Theoretical and Experimental Investigation of the Effects of a Steady Angle of Attack on the Nonlinear Flutter of a Delta Wing Plate Model. *J. Fluids Struct.* **2003**, *17*, 243–259.

15. Tang, D.M.; Dowell, E.H. Theoretical and Experimental Aeroelastic Study for Folding Wing Structures. *J. Aircr.* **2008**,*45*, 1136–1147.

16. Peter, J.A.; Tang, D.M.; Dowell, E.H. Nonlinear Aeroelastic Study for Folding Wing Structures. *AIAA J.* **2010**, *48*, 2187–2195.

17. Tang, D.M.; Henry, J.K.; Dowell, E.H. Nonlinear Aeroelastic Response of a Delta Wing to Periodic Gust. *J. Aircr.* **2000**,*37*, 155–164.

18. Tang, D.M.; Henry, J.K.; Dowell, E.H. Effects of Steady Angle of Attack on Nonlinear Gust Response of a Delta Wing Model. *J. Fluids Struct.* **2002**, *16*, 1093–1110.

19. Tang, D.M.; Dowell, E.H. Experimental and Theoretical Study of Gust Response for a Wing-Store Model with Freeplay. *J. Sound Vib.* **2006**, *295*, 659–684.

20. Conner, M.C.; Tang, D.M.; Dowell, E.H.; Virgin, L.N. Nonlinear Behavior of a Typical Airfoil Section with Control Surface Freeplay: A Numerical and Experimental Study. *J. Fluids Struct.* **1997**, *11*, 89–112.

21. Tang, D.M.; Dowell, E.H.; Virgin, L.N. Limit Cycle Behavior of an Airfoil with a Control Surface. *J. Fluids Struct.* **1998**,*12*, 839–858.

22. Tang, D.M.; Kholodar, D.; Dowell, E.H. Nonlinear Aeroelastic Response of a Typical Airfoil Section with Control Surface Freeplay. In Proceedings of the 41st Structures, Structural Dynamics, and Materials Conference and Exhibit, AIAA 2000-1621, Atlanta, GA, USA, 3–6 April 2000.

23. Tang, D.M.; Dowell, E.H. Aeroelastic Airfoil with Free Play at Angle of Attack with Gust Excition. *AIAA J.* **2010**, *48*, 427–442.

24. Tang, D.M.; Gavin, H.P.; Dowell, E.H. Study of Airfoil Gust Response Alleviation Using an Electro-Magnetic Dry Friction Damper–Part 2: Experiment. *J. Sound Vib.* **2004**, *267*, 875–897.

25. Lee, D.-H.; Chen, P.C.; Tang, D.M.; Dowell, E.H. Nonlinear Gust Response of a Control Surface with Freeplay. In Proceedings of the 51st AIAA/ASME/AHS/ASC Structures, Structural Dynamics and Materials Conference, AIAA 2010-3116, Oriando, FL, USA, 12–15 April 2010.

26. Tang, D.M.; Dowell, E.H. Aeroelastic Response Induced by Freeplay: Part II, Theoretical/Experimental Correlation Analysis. *AIAA J.* **2011**, *49*, 2543–2554.

27. Tang, D.M.; Dowell, E.H. Computational/Experimental Aeroelastic Study For A Horizontal Tail Model with Freeplay.*AIAA J.* **2013**, *51*, 341–352.

28. Tang, D.M.; Dowell, E.H. Effects of a Free-to-Roll Fuselage On Wing Flutter: Theory and Experiment. *AIAA J.* **2014**, *52*, 2625–2632.

29. Tang, D.M.; Dowell, E.H. Experimental Aerodynamic Response for an Oscillating Airfoil in Buffeting Flow. *AIAA J.***2014**, *52*, 1170–1179.

30. Tang, D.M.; Dowell, E.H. Experimental Aeroelastic Response for a Freeplay Control Surface in Buffeting Flow. *AIAA J.***2013**, *51*, 2852–2861.

31. Hodges, D.H.; Dowell, E.H. *Nonlinear Equations of Motion for the Elastic Bending and Torsion of Twisted Nonuniform Rotor Blades*; Technical Report NASA TN D-7818; NASA: Washington, DC, USA, 1974.

32. Tran, C.T.; Petot, D. Semi-Empirical Model for the Dynamic Stall of Airfoils in View to the Application to the Calculation of

Responses of a Helicopter Blade in Forward Flight. *Vertica* **1981**, *5*, 35–53.

33. Simmonds, J.G.; Libal, A. Exact equations for the inextensional deformation of cantilevered plates. *J. Appl. Mech.* **1979**,*46*, 631–636.
34. Simmonds, J.G.; Libal, A. Alternate exact equations for the inextensional deformation of arbitrary, quadrilateral, and triangular plates. *J. Appl. Mech.* **1979**, *46*, 895–900.
35. Simmonds, J.G. Exact equations for the large inextensional motion of plates. *J. Appl. Mech.* **1981**, *48*, 109–112.
36. Darmon, P.; Benson, R.G. Numerical Solution to an Inextensible Plate Theory with Experimental Results. *J. Appl. Mech.* **1986**, *53*, 886–890.
37. Tang, D.M.; Zhao, M.; Dowell, E.H. Inextensible Beam and Plate Theory: Computational Analysis and Comparison with Experiment. *J. Appl. Mech.* **2014**, *81*.
38. Molyneux, W.G. *The Flutter of Swept and Unswept Wing with Fixed-Root Conditions*; HM Stationery Office: London, UK, 1950.
39. Yates, E.C., Jr. *AGARD Standard Aeroelastic Configurations for Dynamic Response I—Wing 445.6*; NASA-TM-100492; NASA Langley Research Center: Hampton, VA, USA, 1 August 1987.
40. Weatin, M.F.; Goes, L.C.; Ramos, R.L.; Silva, R.G. Aeroservoelastic Modeling of a Flexible Wing for Wind Tunnel Flutter Test. In Proceedings of 20th International Congress of Mechanical Engineering (p.9), Gramado, RS, Brasil, 15–20 November 2009.
41. Edwards, J.W.; Schuster, D.M.; Spain, C.V.; Keller, D.F.; Moses, R.M. MAVRIC Flutter Model Transonic Limit Cycle Oscillation Test. In Proceeding of the 19th AIAA Applied Aerodynamics Conference, AIAA-2001-1291, Anaheim, CA, USA, 16–19 April 2001.
42. Peters, D.A.; Cao, W.M. Finite State Induced Flow Models, Part I: Two-Dimensional Thin Airfoil. *J. Aircr.* **1995**, *32*, 313–322.
43. Niles, R.H.; Spielberg, I.N. *Subsonic Flutter Tests of an Unswept All-Movable Horizontal Tail*; WADC TR-54-53, AD 39949;

Wright Air Development Center Wright-Patterson AFB: Dayton, OH, USA; March; 1954.

44. Niles, R.H. *Subsonic Flutter Model Tests of an All-Movable Stabilizer with 35° Sweepback*; WADC Technical Note 55-623, AD 91593; Wright Air Development Center Wright-Patterson AFB: Dayton, OH, USA; November; 1955.
45. Cooley, D.E.; Murphy, J.A. *Subsonic Flutter Model Tests of a Low Aspect Ratio Unswept All-Movable Tail*; WADC-TR-58-31, AD 142337; Wright Air Development Center Wright-Patterson AFB: Dayton, OH, USA; February; 1958.
46. Raveh, D.E. Numerical Study of an Oscillating Airfoil in Transonic Buffeting Flows. *AIAA J.* **2009**, *47*, 505–515.
47. Raveh, D.E.; Dowell, E.H. Frequency Lock-in Phenomenon for Oscillating Airfoils in Buffeting Flows. *J. Fluids Struct.* **2011**, *27*, 89–104.
48. Dowell, E.H.; Hall, K.C.; Thomas, J.P.; Kielb, R.E.; Spiker, M.A.; Denegri, C.M., Jr. A New Solution Method for Unsteady Flows Around Oscillating Bluff Bodies. In IUTAM Symposium on Fluid-Structure Interaction in Ocean Engineering, Proceedings of the IUTAM Symposium, Hamburg, Germany, 23–26 July 2007; Springer: New York, NY, USA; p. 37.
49. Chen, J.; Fang, Y.C. Lock-on of Vortex Shedding due to Rotational Oscillations of a Flat Plane in a Uniform Stream. *J. Fluid Struct.* **1997**, *12*, 779–798.
50. Hall, K.C.; Thomas, J.P.; Clark, W.S. Computation of Unsteady Nonlinear Flows in Cascades Using a Harmonic Balance Technique. *AIAA J.* **2002**, *40*, 879–886.
51. Besem, F.; Kamrass, J.D.; Thomas, J.P.; Tang, D.M.; Kielb, R.E. Vortex-Induced Vibration and Frequency Lock-in of An Airfoil at High Angles of Attack. *J. Fluids Eng. Trans. ASME* **2016**, *138*.

8

CHAPTER

Investigation of Drying Mechanism of Solids Using Wind Tunnel

Abdulaziz Almubarak[1]

[1] *College of Technological Studies, Department of Chemical Engineering, Kuwait*

1. INTRODUCTION

Drying of solids provides a technical challenge due to the presence of complex interactions between the simultaneous processes of heat and mass transfer, both on the surface and within the structure of the materials being dried. Internal moisture flow can occur by a complex mechanism depending on the structure of the solid body, moisture content, temperature and pressure in capillaries and pores. External conditions such as temperature, humidity, pressure, the flow velocity of the drying medium and the area of exposed surface also have a great effect on the mechanisms of drying.

Theoretical and experimental studies [1-6] reported the forced convection heat and mass transfer across different shapes. Most of these studies have considered a heated solid surface with a uniform surface temperature. However, this situation is not the same as in the drying process, where heat and mass transfer occurs simultaneously and the interfacial temperature and moisture content vary during the drying.

Evaporation from a flat plate surface to a laminar boundary layer was theoretically analyzed [7]. It was calculated distributions of the interfacial temperature and local Nusselt and Sherwood numbers for a parallel flow where both Prandtl and Schmidt numbers are unity. A conclusion stated that the characteristics of heat and mass transfer are highly conjugated and significantly influenced by the temperature dependency of vapor-liquid equilibrium, the magnitude of the latent heat of the phase change, and the thermal conductance of the flat plate. The work needs to be extended for the case of unsteady state conditions and to be repeated for a drying bed.

The variation in the surface temperature for a flat bed of a capillary porous material was discussed in [8]. The authors presented graphs that show an increase in surface temperature from the leading edge of the flat plate during the periods of drying. They also showed graphs in which the surface temperature decreased. In general, a physical meaning for that contradiction was needed to be considered.

As in [9], Jomaa *et al.* presented a simulation of the high-temperature drying of a paste in a scaled-up wind tunnel. They attempted to study the influence of the local air flow and thermal radiation on the drying behaviour of the product. A rapid air velocity (8 m s^{-1}) was used in the experiment, and an empirical model was derived to predict temperature and solvent content along the conveyer belt. Comparison of the experimental results with those predicted ones by the model showed a satisfactory agreement.

Evaporation of a pure liquid droplet has been widely studied both theoretically and experimentally [10-14]. However, in spray dryers and droplet drying applications, the droplets always consisted of multi-component mixtures of liquids and sometimes dissolved solids, forming a complicated multiphase composition. This makes analyzing heat and mass transfer processes more difficult. This effect is attributed to the presence of various components that vaporize at different rates, giving rise to a gradient in concentration in the liquid and vapor phase. In addition, a solid crust forms at the outer surface of droplet which acts as a resistance to heat and mass transfer processes.

Various experimental techniques have been used [15-19] to study the mechanisms of drying for a single droplet containing dissolved solids. The droplet was suspended freely from the end of a stable nozzle fixed in a wind tunnel. Air flow was hitting the droplet from one side causing a significant disturbance to the shape of droplet. Therefore, there was some difficulty in recording the weight and temperature of the droplet. The transferred heat conduction to the droplet by the nozzle was another problem.

Cheong *et al.* [20] proposed a mathematical model to predict core temperature for drying a free suspended droplet against time. Reasonable agreement was obtained between the predicted and the experimental results at an air temperature of 20ºC. However, the predicted temperature was less accurate at higher temperatures (50ºC and 70ºC); the model was not applicable for cases at high air temperatures.

A mathematical model was modified in [21], taking into account the droplet shrinkage. The droplet was assumed first to undergo sensible heating with no mass change. The model showed that temperature distribution within the droplet cannot be ignored even for a small diameter droplet of 200 μm.

Wind tunnel definitely is considered one of the best tools to investigate and to study the role of boundary layer and the mechanisms of drying process. The most important variables in any drying process such as air flow, temperature and humidity are usually easy to be controlled inside the wind tunnel. In the current study, through an experimental work and mathematical approach, we attempt to understand the role of the boundary layer on the interface behavior and the drying mechanisms for various materials of a flat plate surface and a single droplet shape.

2. BOUNDARY LAYER OVER A FLAT PLATE SURFACE

A boundary layer developed over a flat-plate surface plays a great role in the mechanisms of convective drying. Very little work has been done on the conjugated problem of heat and mass transfer during a flow over a drying bed. In this paper, through an experimental and mathematical approach, we attempt to understand the role of the thermal boundary layer on the interface behavior and the drying mechanisms for porous mediums. Beds of desert sand, beach sand and glass beads were subjected to forced convective drying in a scaled-up wind tunnel.

A laboratory-scale dryer designed for this work is shown in Figure 1. The apparatus consists of wind tunnel, molecular-sieve air dryer, 3 KW air heater and a fan. Through the wind tunnel a controlled flow of hot, dry air, with an average velocity of 1 m s^{-1}, was passed over a sample mounted flush with the tunnel floor. The last section of the wind tunnel (converging section) was designed to be opened easily for installation of the test bed.

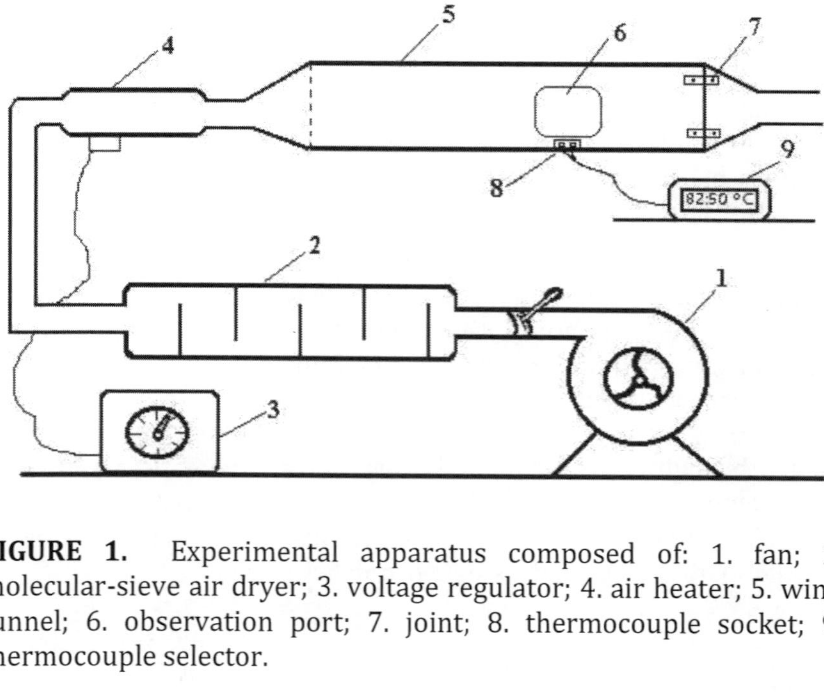

FIGURE 1. Experimental apparatus composed of: 1. fan; 2. molecular-sieve air dryer; 3. voltage regulator; 4. air heater; 5. wind tunnel; 6. observation port; 7. joint; 8. thermocouple socket; 9. thermocouple selector.

A glass tray (100 cm x 36 cm x 2.5 cm) attached to a flat metal plate was especially designed for this study. The top surface of the glass tray was at the same level of the metal plate that formed the floor of the wind tunnel. The sides and the bottom of the tray are insulated with neoprene rubber to minimize heat transfer *via* the glass wall. Figure 2 shows both the wind tunnel and the glass tray.

Ten thermocouples were inserted from one side of the tray to facilitate measurement of surface temperature distribution along the bed. At the same side of the tray, another ten thermocouples were inserted but at lower depth, 2.2 cm from the surface, to measure the bottom temperature distribution.

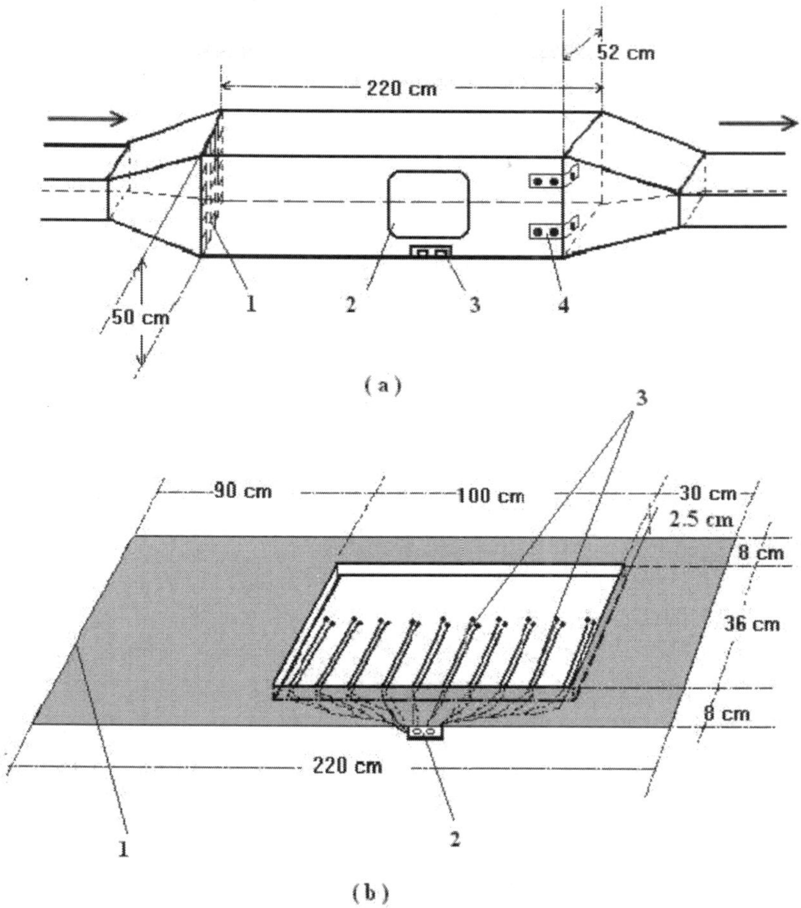

FIGURE 2. Details of (a) Wind tunnel: 1. smoothing grid; 2. observation port; 3.thermocouple socket; 4. joint, (b) Glass tray (clear) and flat plate (gray): 1. the leading edge; 2. thermocouple socket; 3. thermocouples.

The experiment was initiated by switching on the centrifugal fan and then the electric air heater. The voltage regulator was adjusted to provide the desired air temperature. The air temperature was monitored until it reached a steady state. This state was normally required 1 - 2 hours to be achieved.

When the apparatus achieved a constant air temperature, the drying process was commenced. The metal plate and the glass tray, containing the sample, were placed carefully inside the tunnel, allowing the hot air to pass over the surface of the bed. The initial readings of time and temperature were then registered immediately. During the experiment, temperature distributions were measured at intervals of approximately 20 minutes. The temperature at each distance was measured by using the thermocouple selector and registered with an accuracy ± 0.05ºC.

Two types of sand were subjected to the drying process. The first type was a fine sand of average grain size 220 μm diameter, with a moisture content of 0.17 kg kg^{-1}, taken from the desert of Kuwait, Burgan (N 28° 44 00 E 047° 42 00). The nature of the desert sand is usually fine and dry. The other type of sand was a beach sand of average grain size 300 μm diameter and 0.24 kg kg^{-1} moisture content, taken from the north coast of Kuwait, Bobiyan Island (N 29° 46 00 E 048° 22 00). The north coast is a place where the Shatt Al-Arab River (Iraq) falls into the Arabian Gulf. This makes the coast more muddy and less saline than the south coast. In addition, the beach sand contains a lot of small shells of various shapes. Glass beads of 400 μm diameter and 0.2 kg kg^{-1} initial moisture content were also selected, for test as a porous medium for comparison. The bed of glass beads is considered a typical sample for drying experiments, since the sizes of the beads are almost identical. The two types of sand and the glass beads beds are good examples for testing drying processes.

2.1. Results

A bed of the desert sand was subjected to forced convective drying at 84 °C. The wet-bulb temperature of the air was 35.5 °C. The resulting surface temperature vs. time at different distances from the leading is shown in Figure 3, which demonstrates clearly the stages of the drying process; i.e. the pre-constant rate period, the constant rate period, and the falling rate period.

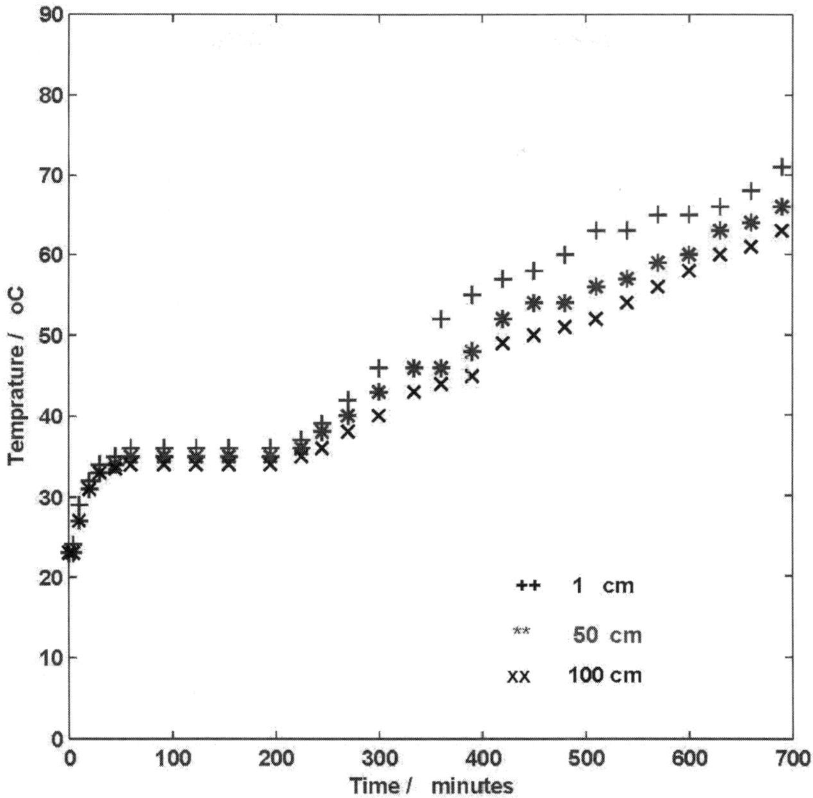

FIGURE 3. Temperature distribution profile of the surface for the desert sand bed.

The temperature of the bed rose from ambient temperature 23°C to a steady value at time = 70 minute. This initial period, termed the pre-constant rate period, is usually short. The surface temperatures remained constant for a period of 140 minutes, indicating the constant rate period. The surface temperature at the distance x=1cm, from the leading edge was 36 °C. This temperature was greater than that at x=50 cm and at x= 100 cm by 1 °C and 2 °C respectively. The surface temperatures were close to the wet-bulb temperature.

During the constant rate period, the surface of the solid is so wet that a continuous film of water exists on the drying surface. This water is entirely unbound water and exerts a vapour pressure equal to that of pure water at the same temperature. The rate of moisture movement within the solid is sufficient to maintain a saturated condition at the surface.

At a specific point, t = 240 minutes, the surface temperature at all positions rose gradually, indicating the end of the constant-rate period and the beginning of the falling-rate period. In the falling-rate period, there was insufficient water on the surface to maintain a continuous film. The entire surface was no longer wetted and dry patches began to form. The surface temperature continued to rise for a longer time until it approximated to the air temperature. A thin dried layer appeared on the entire surface.

The temperatures at the surface and the bottom at different distances from the leading edge are shown in Figure 4. The profiles show that when the surface became dry, the bottom remained wet at a constant temperature for a longer time than that for the surface. During the falling-rate periods a receding evaporation front divided the system into a hotter, dry zone near the surface and a wet zone towards the bottom of the sample [22]. The evaporation plane receded from the surface toward the bottom. The temperatures then rose quickly when the dry zone extended throughout the bed.

A bed of the beach sand was dried at an air temperature of 83 °C. The wet bulb temperature was 35°C. The temperature distribution profile at different distances from the leading edge is shown in Figure 5. The stages of the drying process can easily be recognized from the temperature profile. However, the surface temperature took longer to approach to the air temperature.

Figure 6 shows the surface and the bottom temperatures at distances from the leading edge of 1 cm, 50 cm 100 cm, respectively. It can be seen that the surface and bottom temperatures increased rapidly at some times and decreased others. Also, the temperature profile did not increase gradually like that of the desert sand. This can be attributed to the nature of the beach sand, which comprises different types of small shells of various shapes, and contain tiny hollows. The trapped water in these hollows forms small bubbles which can explode with increasing temperature. Therefore the temperature of the bed changes suddenly at such times. The temperature at the bottom of the bed indicated it remained wet during the falling-rate period.

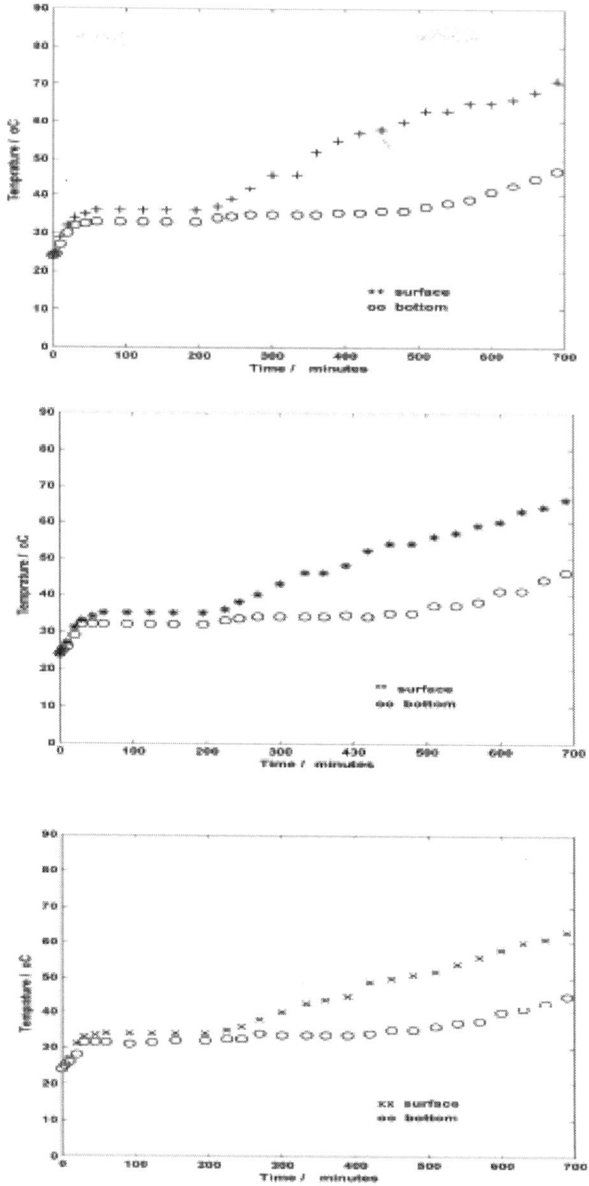

FIGURE 4. Temperatures of surface and bottom for the desert sand bed at distance of 1 cm, 50 cm and 100 cm from the leading edge, respectively, from the top.

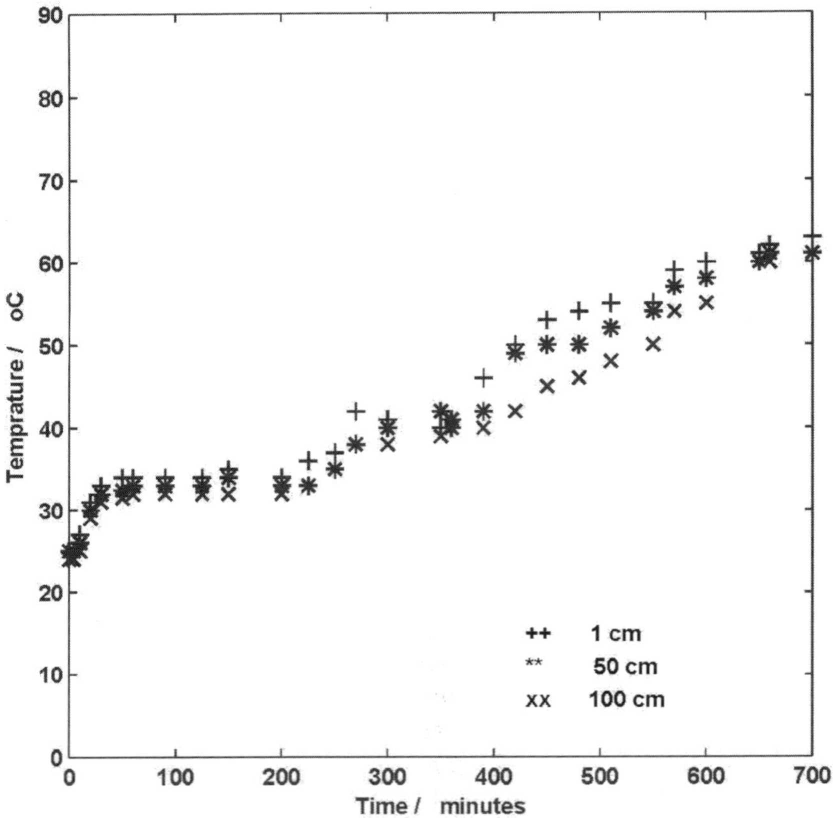

FIGURE 5. Temperature distribution profile of the surface for the beach sand bed).

A similar investigation was carried out for a bed of the glass beads at an air temperature of 84 °C. The wet bulb temperature was 37 °C. Figure 7 shows temperature versus time at different distances from the leading edge of the bed surface. The temperature distribution profile again illustrates clearly the stages of the drying process. The results for the surface and bottom temperatures for the glass beads were almost similar to those for the desert sand as shown in Figure 4.

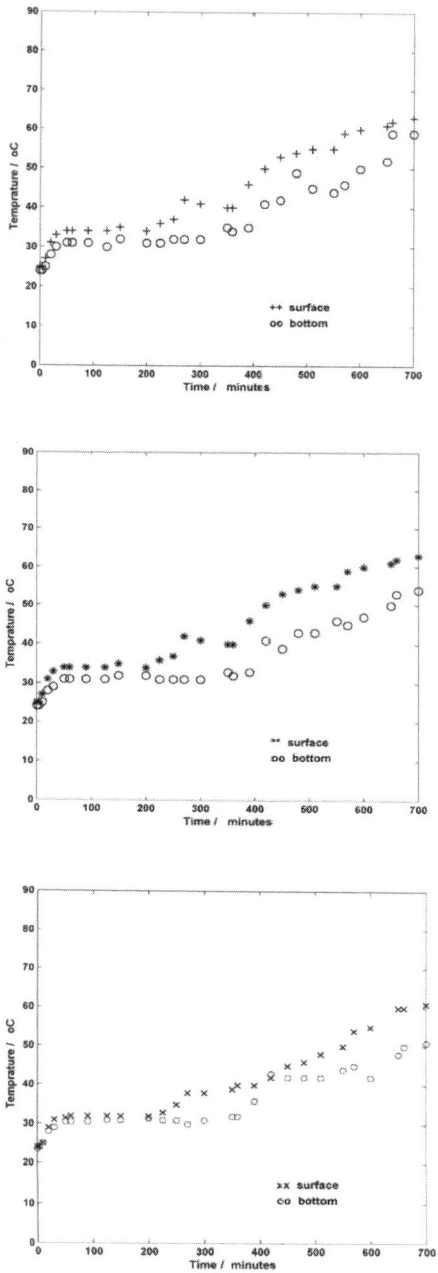

FIGURE 6. Temperatures of surface and bottom for the beach sand bed at distance of 1 cm, 50 cm and 100 cm from the leading edge, respectively, from the top.

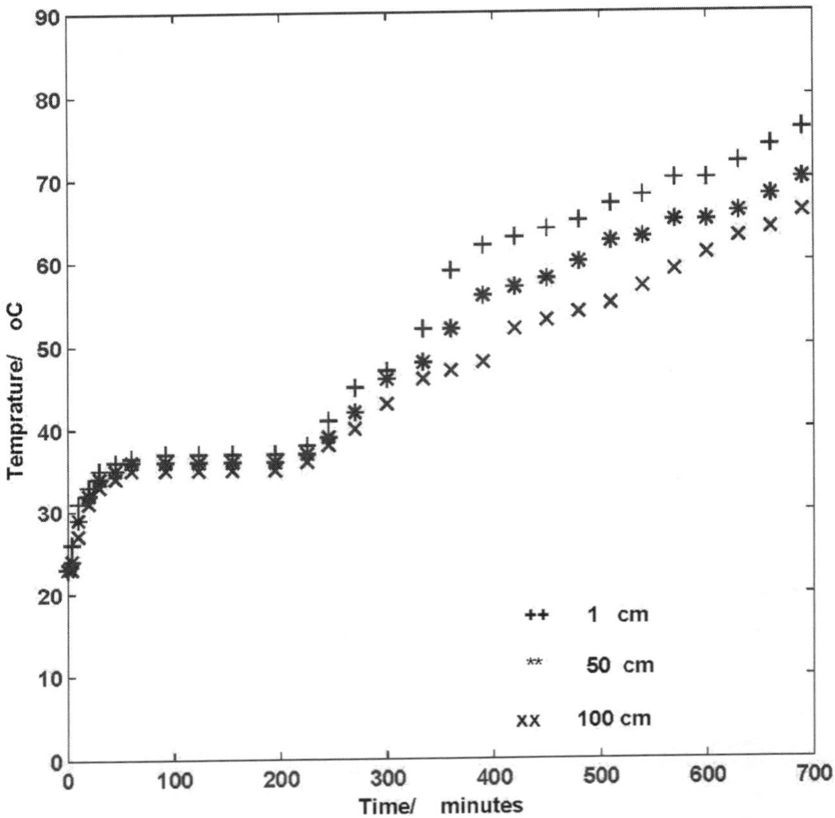

FIGURE 7. Temperature distribution profile of the surface for the glass-bead bed.

2.2. Discussion

A thin film adjacent to the surface always exists when a forced flow passes over a flat plate and forms what is called a hydrodynamic boundary layer. The influence of the surface temperature reaches deeper into the fluid, thus causing the formation of a thermal boundary layer. It is well known, the thickness of the thermal boundary layer increases with increasing distance from the leading edge. This layer is affected by the geometry of the system, roughness of the surface and the fluid properties.

For the case in which the heated section is preceded by an unheated straight length, the local Nusselt number (Nu_x) is represented in [23-25] as:

$$Nu_x = \frac{hx}{k} = \frac{0.323(Re_x)^{0.5}(Pr)^{1/3}}{[1-(x_o/x)^{3/4}]^{1/3}} \tag{1}$$

where *Rex* is Reynolds number with respect to length and *x* is the length of the flat plate in (m).

Figure 8 shows a plot of variation in the local heat transfer coefficient versus the distance from the leading edge. The plot indicates that the values of the coefficient decreased significantly when x increased from the leading edge, and then it remains virtually constant for large x values. A plot of variation of the mass transfer coefficient could be expected to be almost similar to that in Figure 8because of the similarity in the transport coefficient equations. This result with the concept of the boundary layer thickness demonstrates that resistance to heat and mass transfer to, or from, the bed increases with increasing the distance from the leading edge.

A model proposed in a previous paper [26] was modified to find a method for prediction of surface temperature distribution. The equation of energy can be represented as:

$$\varrho\, cp \frac{\partial T}{\partial t} = \nabla(k\nabla T) + \nabla(\frac{D_v M}{RT}\nabla P_v)\Delta hv \tag{2}$$

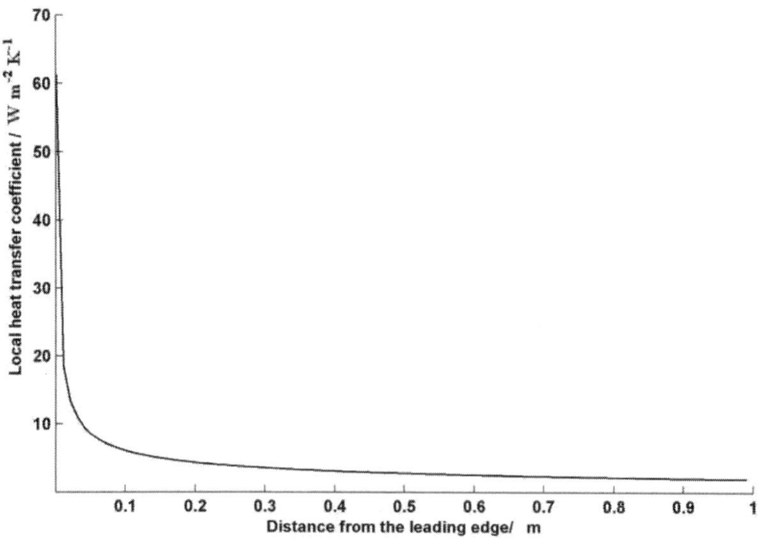

FIGURE 8. Local heat transfer coefficient vs. distance from the leading edge.

At time zero, the whole body has a uniform initial temperature of T_o, and the initial conditions are:

$$T_o = T\big|_{y=0} = T\big|_{y=h} \tag{3}$$

At the external surface, i.e. y=0, the boundary conditions can be written on the basis of Figure 9 as:

$$-k\,\frac{\partial T}{\partial y} = h\,(T_a - T_{sx}) + \dot{m}\,\Delta hv \tag{4}$$

where T_{sx} is the surface temperature at distance x from the leading edge.

The drying rate, $\dot{m}$ · varies with the time and can be defined as,

$$\dot{m} = \frac{M K'_G}{RT}\left(P_{sr} - P_v\right) \tag{5}$$

where

$K'G$ is an overall mass transfer coefficient, defined in [27] as:

$$\frac{1}{K'_G} = \frac{1}{k'_c} + \frac{z''}{D_{eff}} \tag{6}$$

where $k'c$ is the local mass transfer coefficient (m s^{-1}), z'' is the distance from the plate surface to the receding evaporation front in (m).

The boundary condition at y = h is:

$$-k \frac{\partial T}{\partial y} = 0 \tag{7}$$

Equation (2), with such boundary conditions, was solved by a finite difference method (a modified form the so-called explicit method). Therefore, the temperature distribution on the surface (i.e. y = 0) was calculated by using the model at different local points, x. Table 1 shows the physical properties of the sample materials.

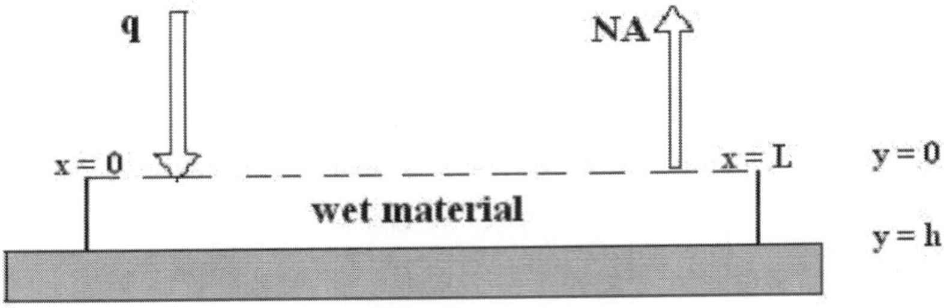

FIGURE 9. Heat and mass transfer process for the wet material. NA: mass transfer flux and q: heat transfer flux.)

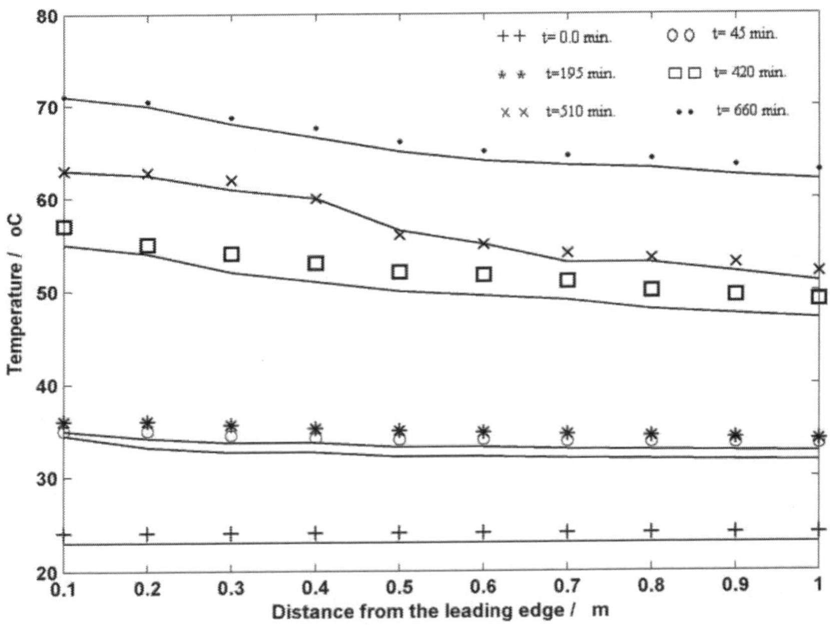

FIGURE 10. Experimental and predicted surface temperatures of the desert sand bed; experimental results (symbols); predicted results (solid lines).

Figure 10 shows the experimental and the predicted surface temperature distributions for the desert sand along the bed at various times. The predicted temperatures were in good agreement with the experimental results. From the graph, it can be seen that there is a significant difference in the surface temperature between

0.1 m and 1 m. At time=195 minutes; i.e. during the constant-rate period, the difference in temperature was 2 °C.

During the falling-rate period, the difference in the surface temperature between 0.1 m and 1 m increased. At 510 minutes, the difference is 11 °C. At the bottom of the sand bed, the difference in temperature along the bed also can be seen clearly (Figure 4).

For the entire time of the experiment, the surface and the bottom temperatures decreased gradually with increasing distance from the leading edge. This caused by the resistance to heat transfer process which increased with increasing thickness of thermal boundary layer. In contrast, near the leading edge, the resistance to heat transfer diminishes, since the thickness of the thermal boundary layer in the vicinity of the surface thins. Therefore, the rate of heat transfer to the body increased, thereby raising the temperature of surface. Afterwards, heat transfer by conduction across the solid particles raises the temperature of the bed, and the portion closest to the leading edge dries faster than that at a greater distance.

Figure 11 shows both predicted and experimental results for the bed of beach sand. This Figure shows that the computed temperature distribution was in general agreement with experimental results. However, unsatisfactory results can be seen at times greater than 400 minutes, i.e. during the falling-rate period. This is due to the nature of the beach sand as discussed before. The surface temperature of the beach sand sample decreased gradually with increasing distance from the leading edge. A significant difference in temperature between 0.1 m and 1 m can be seen clearly for both constant and falling-rate periods. During the constant-rate period, the difference in temperature was 2 °C, whereas in the falling-rate period the difference reached to 8 °C.

We have found that the temperature distribution profiles determined for the flat beds of desert sand, beach sand and glass beads identified clearly the stages of drying. The temperature profiles in general were almost similar. However, the beach sand profile showed irregularity in temperature, due to the nature of the beach sand which contains a lot of shells with various shapes.

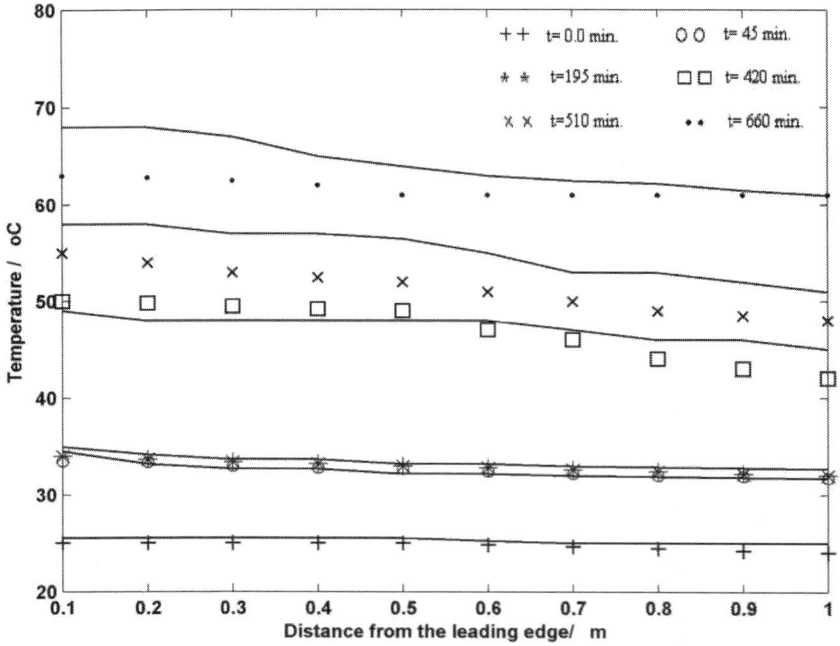

FIGURE 11. Experimental and predicted surface temperatures of the beach sand bed; experimental results (symbols); predicted results (solid lines).

TABLE 1. Physical properties for desert sand and beach sand.

	ρ(Kg/m³)	cp_d (kJ/kg K)	k (W/m K)	k_d (W/m K)	D_{eff} (m²/s)	D_v (m²/s)
Desert sand	1760	1.1	0.760	0.246	0.81×10^{-5}	1.75×10^{-5}
Beach sand	1980	1.2	1.12	0.188	1.15×10^{-5}	2.05×10^{-5}

The temperature profiles also showed that when the whole surface of the bed became dry, the whole bottom of the bed remained wet. During the falling-rate periods, a receding evaporation front divided the system into a hotter, dry zone near the surface and a wet zone towards the bottom of the sample.

The predicted transport coefficients have very large values close to the leading edge, where the thickness of the boundary layer approaches zero. In contrast, the values of the coefficients decrease progressively with increased distance from the leading edge, where the boundary layer thickens. Hence the resistance to heat and mass transfer to, or from, the surface also increases. These variations in thickness and resistance have a significant effect on the temperature distribution along the bed and the drying rate.

A mathematical model has been modified to predict temperature distributions along the bed at various times. The model was compared with the experimental results for various beds and good agreement was obtained. We found that surface and bottom temperatures decreased gradually with increasing distance from the leading edge, and the difference in temperature became clearer during the falling-rate period. The difference in the surface temperature was 11 °C for the case of desert sand, and was 8°C for the case of beach sand. We concluded that the portion close to the leading edge dried faster than that at larger distance, since the resistance to heat and mass transfer diminishes at that position.

3. SINGLE DROPLET DRYING

The study of mechanisms that describe single droplet drying is a challenging issue since it involves many disciplines: heat and mass transfer, fluid mechanics and chemical kinetics. The main objective of this section is to attempt to further understand the mechanisms involved in the drying of a single droplet, and more specifically to formulate a mathematical model. The model should predict temperature profiles for both the inner core and the outer surface of the droplet under simulated conditions that might be encountered in spray drying equipment. Experimental work will also involve the measuring the moisture contents against time to provide more information for the droplet drying process.

The experimental apparatus was comprised of a horizontal wind tunnel 2.2 m long. The wind tunnel supplied a forced drying air into the working section where the droplet was suspended from a glass nozzle. The apparatus and the flow system are shown in Figure 12. A gate valve at the inlet to the wind tunnel controlled airflow rate. Air was heated to the desired temperature, using a 3 KW electric heating element controlled by a rotary voltage regulator.

Through the wind tunnel, a controlled flow of hot, dry air, with an average velocity of 1 m/s, was passed across the droplet suspended from the glass nozzle. A forced dry air was obtained by using a centrifugal fan and a molecular sieve air dryer containing silica gel and calcium silicate.

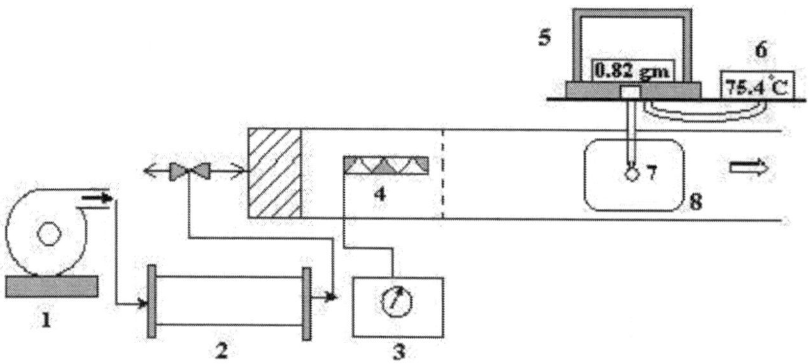

FIGURE 12. Experimental apparatus: 1-fan 2-molecular sieve 3-voltage regulator 4-air heater 5-Digital balance 6-Temperature recorder 7-Glass nozzle 8-Observation port.

The experiment was initiated by switching on the centrifugal fan and then the electric air heater. The voltage regulator was adjusted to provide the desired air temperature. A thermocouple was fixed in the wind tunnel near to the glass nozzle to measure the air temperature with an accuracy of ± 1° C. The air temperature was monitored until it reached a steady state. This state requires between 1 to 2 hours to be achieved. When the apparatus achieved a constant air temperature, the drying process was initiated.

As a part of this study, a droplet suspension device was specially designed to measure the weight and temperature of the droplet. The droplet suspension device is illustrated in Figure 13. It consisted of a glass nozzle with the dimensions of 180 mm in length and 9 mm outside diameter. The upper section of the glass nozzle was fixed by a small electric rotator device able to provide a rotational speed range of 1- 30 rpm. In order to reduce the contact area between the free end of the glass nozzle and the droplet surface, the lower section of the nozzle was shaped as a small cone with a free end diameter of 4 mm. The droplet receives a relatively small amount of heat transferred by conduction through the glass nozzle, which has a thermal conductivity of 0.480 W/m K. The heat transferred by conduction was taken into account in the proposed model represented by Equations 4&5. The glass nozzle was rotated at constant low speed of 5 rpm. This rotation speed had no effect on the shape or stability of the droplet. The upper section of the suspension device was fixed by a metallic clip installed beneath the analytical balance. The lower section of the suspension device, i.e. the glass nozzle, was inserted through a hole in the wind tunnel.

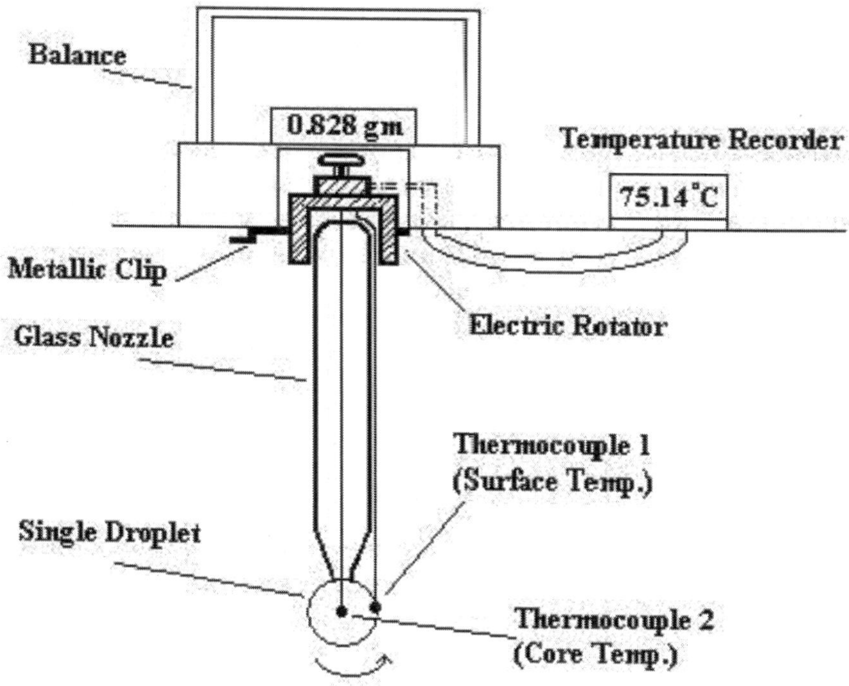

FIGURE 13. Droplet suspension device; glass nozzle and the connected thermocouples.

Two thermocouple sensor types, NiCr-NiAl, were used to measure the core and surface temperatures of the droplet. One of the thermocouples was placed inside the glass nozzle and extended to the center of the droplet. The other thermocouple was fixed outside and along the glass nozzle. The outer thermocouple rotates simultaneously with the rotation of the glass nozzle. The end tip of the thermocouple was positioned in a manner to touch the outer surface of the droplet. The rotation process that made both the nozzle and the droplet rotate together assisted in avoiding any separation between them that might have been caused by the force of air drying. The core and the surface temperatures of the droplet were easily recorded by a temperature recorder at 50 sec. intervals. The drying process of the different material droplets was investigated under air temperature of 75° C and 140° C.

The procedure for weighing the droplet was carried out quickly and intermittently by causing the suspension device to be freely-suspended. The gate valve was closed to cut off the airflow to the working section and diverted to an outlet 20 mm valve in order to prevent any vibration of the glass nozzle during the weighing process. The nozzle rotation was simultaneously stopped. A metallic clip was opened manually to allow the suspension device to be freely-suspended from a hook connected beneath the balance. This arrangement made the weight measurement readings more accurate. The weighing procedure was repeated during the droplet drying experiment at 100 secintervals. Thus, the weight loss of droplet was recorded and moisture content was determined versus time. The required time for each weighing procedure step was about 10 sec. This time was not included in the recorded drying time and had no noticeable effect on the results.

Three types of liquids were selected for the drying process experiments. The first type was sodium sulphate decahydrate solution (60 wt % solid). The second type was a concentrated fruit juice (60 wt % fruit juice powder of apple, peach and blueberry, MTC product). The third type was an organic paste (20% sodium chloride, 25% dispersal pigment; Goteks product) used for adhesive and coating applications. Table 2 shows the physical properties of the sample materials. Droplets ranging from 9 mm to 14 mm diameter were subjected to the drying process. However, the actual size of the droplets in typical spray drying applications is much smaller. The droplets with small sizes require developing a more accurate technique to measure both core and surface temperatures. Therefore, the current research assumes that the mechanisms of the drying large and small droplets are similar.

TABLE 2. Physical properties of sodium sulphate decahydrate solution, fruit juice and organic paste.

	ρ_d (Kg/m^3)	cp_d (kJ/kgK)	k (W/mK)	k_d (W/mK)	D_{eff} (m^2/s)	D_v(m^2/s)
Sodium sulphate decahydrate solution	3110	1.1	0.180	0.246	1.14×10^{-5}	3.45×10^{-5}
Fruit juice	1650	1.7	0.126	0.188	1.10×10^{-5}	3.40×10^{-5}
Organic paste	3400	2.4	0.251	0.422	1.52×10^{-5}	3.50×10^{-5}

3.1. Mathematical Model

In the drying process, the droplet is first heated by the hot air flow with significant evaporation from the surface. The temperature of the surface increases and approaches the wet-bulb temperature, indicating the constant drying rate period. In order to propose a mathematical model, the droplet was assumed to have a fixed size with no change during this period. Also, the droplet was assumed to have a uniform initial temperature and moisture content. Temperature distribution within the droplet can be represented as

$$\frac{\partial T}{\partial t} = \alpha\left(\frac{\partial^2 T}{\partial r^2} + \frac{2}{r}\frac{\partial T}{\partial r}\right) \tag{8}$$

Equation 8 was solved with the following boundary conditions using explicit finite differences,

$$-k\frac{\partial T}{\partial r} = 0 \qquad at\ r = 0 \tag{9}$$

and

$$-k\frac{\partial T}{\partial r} = h(T_a - T_{sr}) + q_{nz} \qquad at\ r = R \tag{10}$$

where, q_{nz} is the transferred heat conduction to the droplet through the glass nozzle. It can be calculated as:

$$q_{nz} = \left(\frac{4 k_{nz} h_{nz}}{d_{nz}}\right)^{0.5} (T_a - T_{sr}) \tag{11}$$

The heat transfer coefficient, h_{nz}, can be correlated from Thomas (1999) as:

$$h_{nz} = 0.26 + Re^{0.6} Pr^{0.33} \left(k_{air} / d_{nz}\right) \tag{12}$$

During the falling rate period, the formation of a receding evaporation front divides the droplet into two regions, a dry crust at the outer surface and a wet region inside the core. Therefore, heat transfer equations are formulated for each region. The physical model and the coordinate system for analysis are shown in Figure 14.

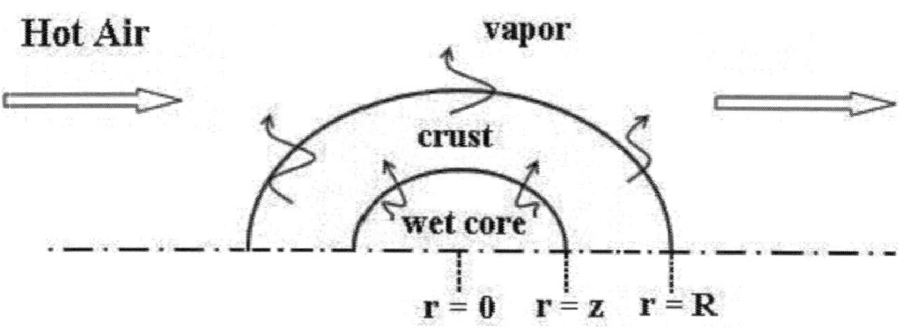

FIGURE 14. Physical model and the coordinate system of the droplet cross-section.

Energy balance for the wet core, 0 < r < z, can be represented as follows:

$$\frac{\partial T_w}{\partial t} = \alpha_w \left(\frac{\partial^2 T_w}{\partial r^2} + \frac{2}{r} \frac{\partial T_w}{\partial r} \right) \qquad (13)$$

Heat is transferred through the crust into the wet core where evaporation occurs at the interface between the core and the crust. Vapor then diffuses through pores of the crust into the drying medium. Thus, moisture is transferred mainly by vapor flow. Consequently, vapor diffusion must be taken into account in formulating the equations for the dry (crust) region. The energy balance for the crust region, z < r < R, can be represented as follows:

$$\frac{\partial T_d}{\partial t} = \alpha_d \left(\frac{\partial^2 T_d}{\partial r^2} + \frac{2}{r} \frac{\partial T_d}{\partial r} \right) + \frac{\partial}{\partial r} \left(\frac{M}{RT} \frac{\partial P_v}{\partial r} \right) \frac{\Delta h_v}{\rho_d c p_d} D_v \qquad (14)$$

3.2. BOUNDARY CONDITIONS

At the center of the sphere, r = 0

$$-k_w \frac{\partial T_w}{\partial r} = 0 \qquad (15)$$

At the surface, r = R:

$$-k_d \frac{\partial T_d}{\partial r} = h(T_a - T_{sr}) + q_{nz} \qquad (16)$$

At the receding evaporation front (r = z), the moving boundary conditions are

$$-k_d \frac{\partial T_d}{\partial r} + k_w \frac{\partial T_w}{\partial r} = \dot{m}\, \Delta hv \tag{17}$$

The drying rate, $\dot{m}$, was defined in Eqs. (5 & 6). However, in this case, z″ , in Eq. (6) represents the distance from the droplet surface (r = R) to the receding evaporation front (r = z).

Heat and mass transfer coefficients, h and k′c , can be determined by the correlations found in [28,29] as follows:

$$Nu = 2 + \Phi Re^{0.5} Pr^{0.33} \tag{18}$$

and

$$Sh = 2 + \beta Re^{0.5} Sc^{0.33} \tag{19}$$

where Φ and β are constants ranging from 0.6-0.7 for Re (500 - 17000).

Thermal conductivity of the wet region can be evaluated according to [30] as:

$$k_w = k_d + k_v \overline{X} \tag{20}$$

where $\overline{X}$ is an average moisture content and kv

is

$$k_v = \frac{D_v M}{RT} \cdot \frac{dp}{dT} \Delta h_v \tag{21}$$

The non-linear Eqs. (13&14) with the boundary conditions as in Eqs. (15-17) were solved by a finite difference method. The proposed equations were solved in a program using Turbo-Pascal V.6, and the computational results compared with the experimental results.

3.3. Results

A single droplet of sodium sulphate decahydrate solution was suspended from the free end of the glass nozzle at an air temperature of 75° C and an air velocity of 1 m/s. The wet-bulb temperature was 46°C. The core and surface temperatures of the droplet versus time are plotted in Figure 15. Initially, the temperatures increased rapidly because of the large difference in temperature between the drying medium and the droplet. Basically, the plot exhibited two drying periods. A short period, from time =100 - 200 s, where the surface temperature approached the wet-bulb temperature, is described as the constant rate period. The second period, or falling rate period, extended from t =200 s to the end of the experiment.

Figure 15 plots the predicted values of both the core and surface temperature evaluated by the developed model. Comparison of the experimental and theoretical results showed good agreement. However, actual surface temperatures were slightly higher than those predicted by the model. This probably was due to the position of the thermocouple and its reading during the experiment, as will be discussed later in more detail.

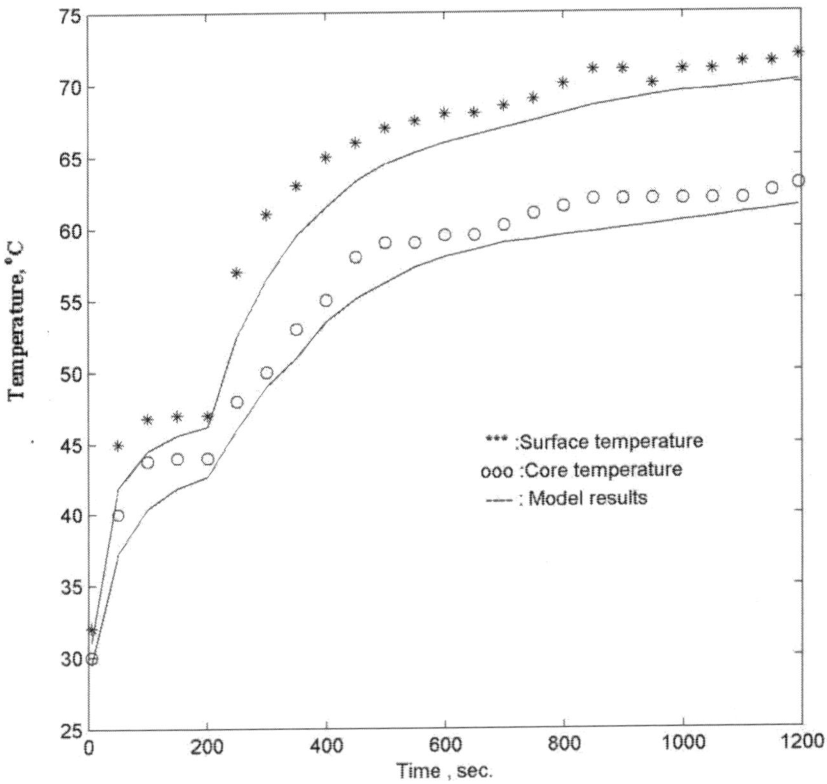

FIGURE 15. Temperature distribution profile for sodium sulphate decahydrate at air temperature of 75° C.

The previous experiment was repeated at an air temperature of 140° C in the same drying medium.Figure 16 shows experimental and theoretical results of a droplet of the same material. The experimental results showed that the constant rate period extended from time= 85 - 165 s, shorter than that observed at 75° C. The results showed both surface and core temperatures of 130° C at 900 s, at which point the droplet had dried completely. Figure 16 showed agreement between the experimental and predicted results.

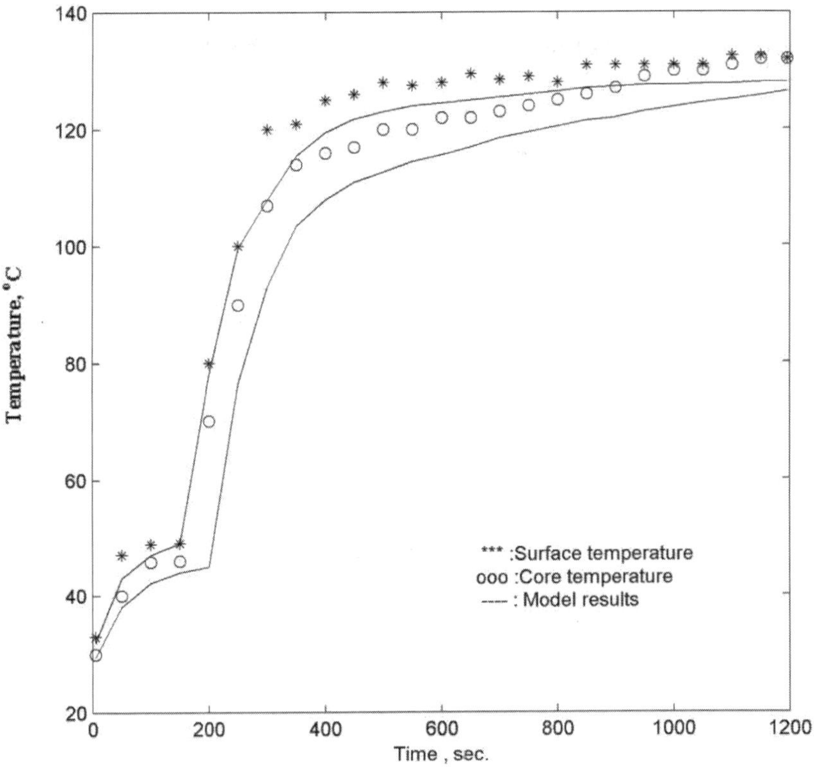

FIGURE 16. Temperature distribution profile for sodium sulphate decahydrate at air temperature of 140° C.

A single droplet of a fruit juice was dried at an air temperature of 75° C and an air velocity of 1 m/s. Wet-bulb temperature was 44.5° C. The experimental results in Figure 17 show a constant rate period from time =100 - 200 s. The falling rate period can be observed after the constant period, when the temperature increased quickly.

During the falling rate period some experimental readings represented approximate values of the actual droplet surface temperatures. The approximate values can be attributed to the fact that the droplet diameter constantly decreased with time and that caused the distance between the tip of thermocouple and the surface of droplet to grow. In other words, during some time in the experiment, a part of the end tip of the thermocouple was touching the surface of the droplet, and the remaining area of the tip was exposed to air flow. Such a behavior caused the tip of thermocouple to give an average reading for both the surface and the air flow temperatures. Good agreement was obtained between the theoretical and experimental results of the temperature profile for the fruit juice.

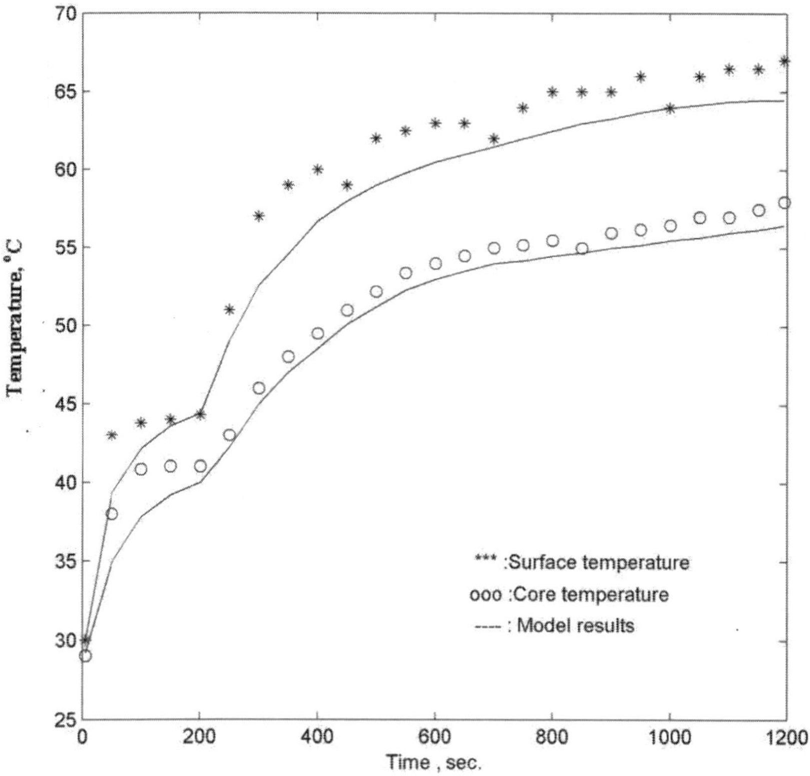

FIGURE 17. Temperature distribution profile for fruit juice at air temperature of 75° C.

The fruit juice droplet was dried again at an air temperature of 140° C in the same drying medium. The results (Figure 18) showed that the constant rate period was a little shorter than that at 75° C. The results showed also that both surface and core temperatures were similar and close to the air temperature at time = 1050 s. The mathematical model showed also good agreement with the experimental readings.

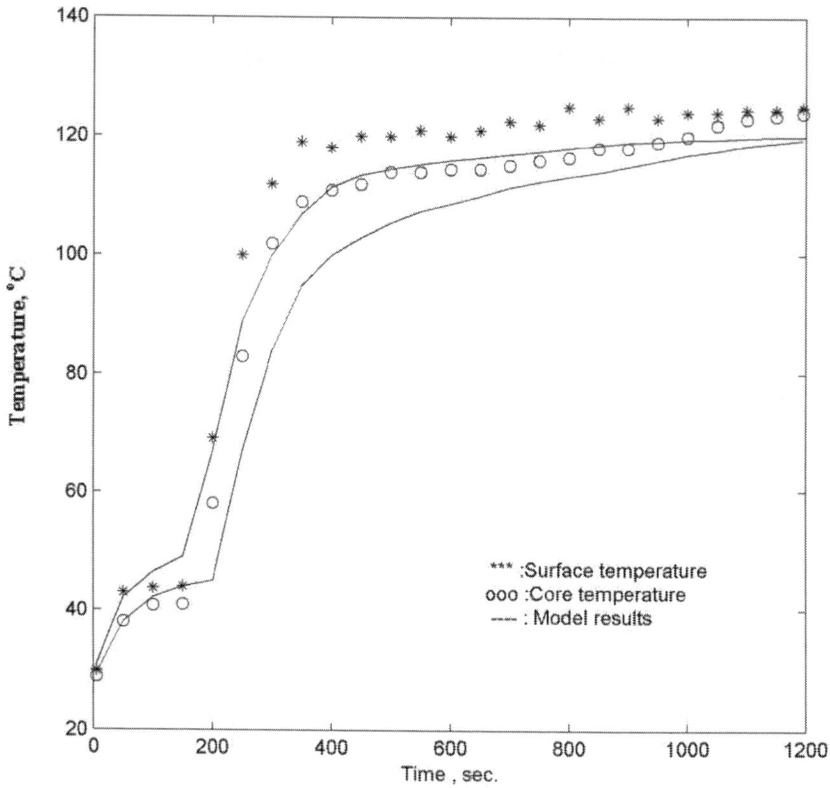

FIGURE 18. Temperature distribution profile for fruit juice at air temperature of 140° C.

A single droplet of organic paste was dried under similar conditions to that of the fruit juice. A plot of the core and surface temperatures of the droplet versus time is shown in Figure 19. A relatively longer period of constant drying rate, compared to those shown for sodium sulphate and fruit juice, was observed. Figure 19 also showed that surface temperature increased rapidly after time = 300 s, indicating the beginning of the falling rate period. Less agreement was obtained between the model's predicted results and the experimental results. Organic paste, which has higher thermal conductivity, forms a thicker solid crust at the outer surface compared to the other material evaluated. Therefore, a higher resistance to heat and vapor through the crust would be expected. To improve the predictions, introduction of a correction factor is probably required in the model for those materials which have a nature similar to that of the organic paste. Figure 20 shows the experimental and theoretical results of air temperature of 140° C. Again, forming a solid crust at the outer surface led to less agreement between theoretical and experimental results. The predicted temperature distributions were higher than the actual temperatures.

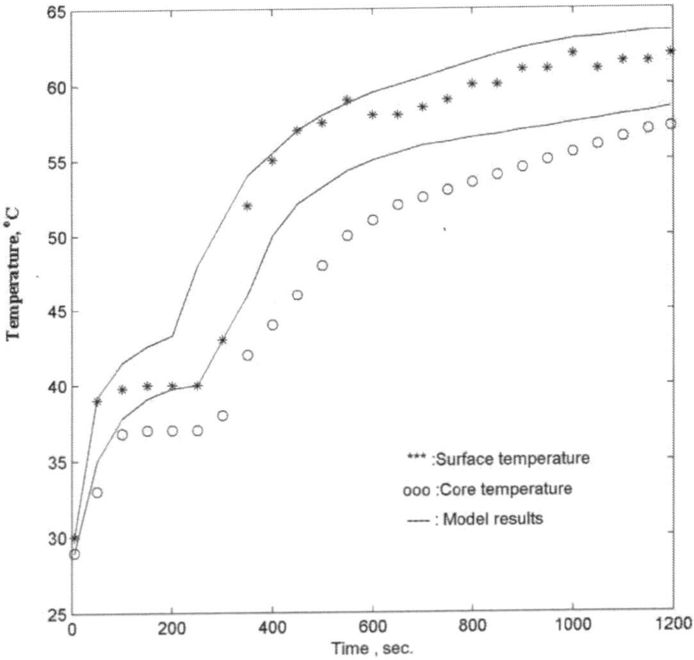

FIGURE 19. Temperature distribution profile for organic paste at air temperature of 75° C.

Obviously, crust formation, thickness and porosity have a significant effect on the moisture content and on the drying rate of the droplet. This result was also obtained by Hayder & Mumford [31] in the drying of custard and starch droplets. They observed that crust formation was more rapid on the custard droplet, because the smaller starch granules absorbed less and left more free water in the droplet. Therefore, the crust growth and the drying rate of the starch droplets took a longer time.

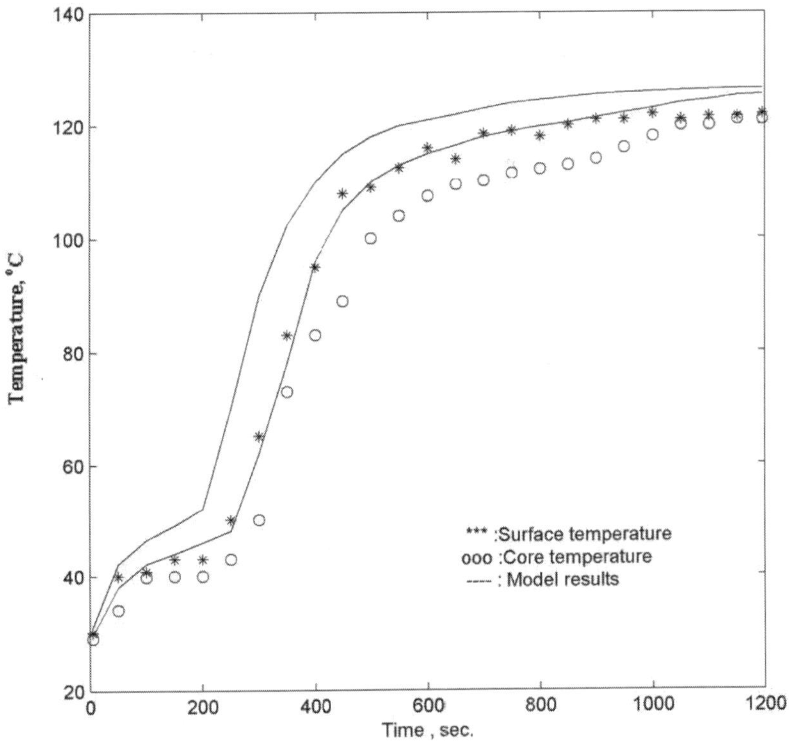

FIGURE 20. Temperature distribution profile for organic paste at air temperature of 140° C.

The moisture content of the droplet was determined by measuring the weight loss against time. There was no experimental technique to measure core and surface moisture content separately; therefore, the measured moisture content of the droplet represented the average value. The experimental results of moisture content distribution for the three samples at an air temperature of 75° C, are shown in Figure 21. The moisture content profiles clearly show the two stages of drying, the constant rate period and the falling rate period. The profiles show that the sodium sulphate decarbohydrate solution had consistently lower moisture content, as it dried faster than the other materials. The profiles also showed that forming a solid crust in the falling rate period lowered the moisture content values for all the three samples. In the case of organic paste, the change in the moisture content was more significant.

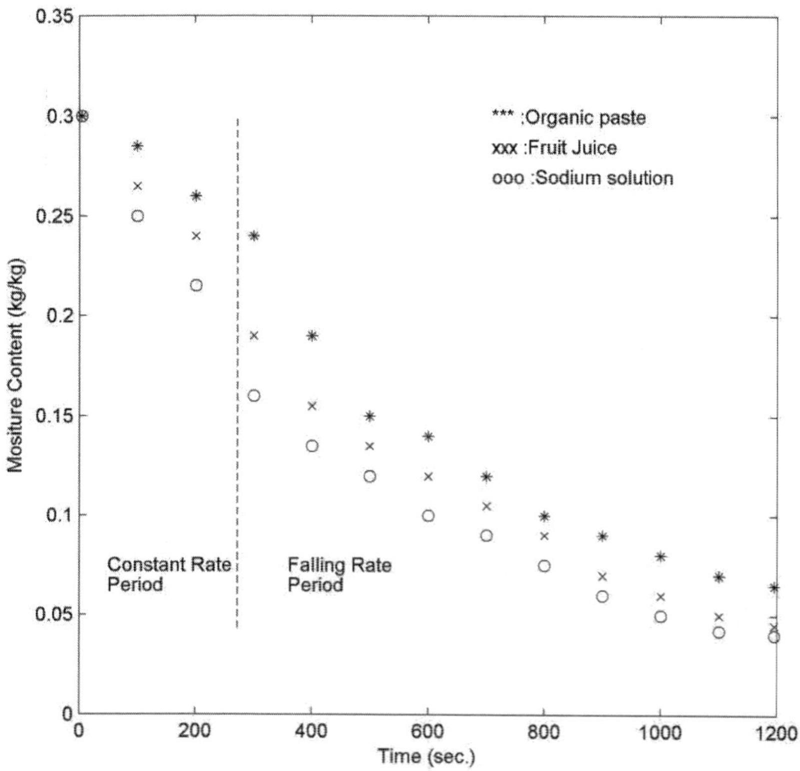

FIGURE 21. Moisture content profile for all samples at air temperature of 75° C.

Moisture distribution curves for the three samples at 140° C air temperature are plotted in Figure 22. Similar results were obtained to those observed at an air temperature of 75° C. Also, the profiles showed that the moisture content dropped to lower values compared to those at 75° C. In other words, the droplets dried faster at the higher temperature.

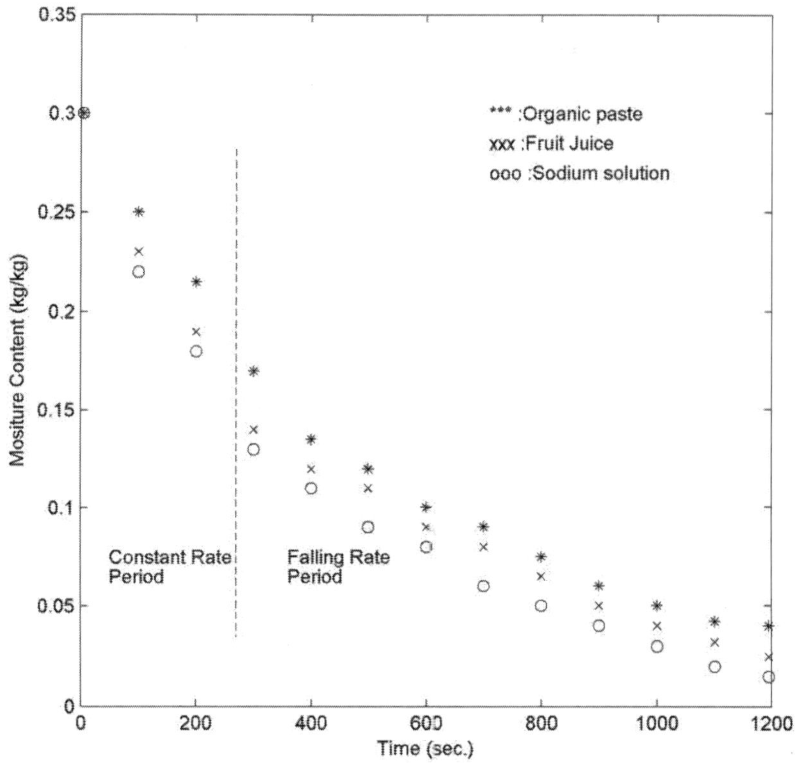

FIGURE 22. Moisture content profile for all samples at air temperature of 140° C.

3.4. Discussion

As previously observed, moisture and temperature distribution profiles of various materials exhibited constant and falling drying rate periods. In the constant rate period, the temperature of the droplet surface was almost equal to the wet-bulb temperature. During this period, evaporation takes place from the free liquid surface of the droplet. The constant rate period was relatively short in sodium sulphate and fruit juice samples. However, that period was longer in the case of organic paste.

The falling rate period is characterized by formation of a partial crust on the outer surface of the droplet. This crust recedes towards the core and the surface temperature starts to increase. Vapor diffusion becomes the predominant transport process at this stage.

Crust structure, thickness and porosity have a significant effect on the rate of drying. The crust thickness increases with time, hence the resistance to heat and moisture diffusion through the crust increases. Therefore, moisture content and drying rate decrease.

Some experimental readings represented approximate values of the actual droplet surface temperatures. This was attributed to the end tip of thermocouple that was giving average readings for both the surface and the air flow temperatures at the same time. It was also assumed that the mechanism of the droplet drying is similar for both small and large diameters. Therefore, it is recommended that a more accurate technique for measuring droplets with small sizes and for the taking of surface temperatures be developed.

The new model predicted temperature distribution profiles for single droplets of various materials. The predicted results showed a good agreement with the experimental data for air temperatures at 75° C and 140° C. However, the model was less accurate in the case of organic paste due to the higher thermal conductivity of the formed crust. A correction factor should be developed and taken into account for such materials. The model provides a relatively fast and efficient way to simulate drying behavior over a range of drying conditions. The model also represents a useful tool in the design and optimization of spray drying processes.

Moisture content profiles clearly showed the two stages of the drying process. In addition, the moisture profiles supported the conclusions that the crust forming in the falling rate period decreased both the moisture content and the drying rate.

Through the results of the experimental work and the theoretical model for a droplet drying, a significant conclusion was obtained. It has always been wrongly assumed in the literature that there is no temperature distribution within the droplet. This concept has been corrected by the current research. All experiments for the three materials used showed a clear difference between the core and the surface temperatures of the droplet during the drying process.

3.5. General Conclusion

The characteristics of the boundary layer have a great effect on the local heat and mass transfer coefficients and temperature distributions throughout a flat bed surface. Droplet drying is an important subject in drying science, since it provides more details about the drying mechanisms in order to optimize spray dryer equipment. However, this part of the drying field is rarely considered in the literature. Further research in this area seems essential to obtain better understanding of drying theory.

Wind tunnel definitely is considered one of the best tools to investigate and to study the mechanisms of drying process. The most important variables in any drying process such as air flow, temperature and humidity are usually easy to be controlled inside the wind tunnel. Through a mathematical approach and an experimental work using a wind tunnel, we highlighted on the role of the boundary layer on the interface behavior and the drying mechanisms for various materials of a flat plate surface and a single droplet shape.

REFERENCES

1. Sparrow E, Lin S. Boundary Layer With Prescribed Heat Flux Application to Simultaneous Convective and Radiation. International Journal of Heat and Mass Transfer 1965; 8, 437-448.

2. Luikov A. Conjugate Convective Heat Transfer Problems. International Journal of Heat and Mass Transfer 1974; 17, 257-265.

3. Chyou T. The effect of A Short Unheated Length and A Concentrated Heat Source on The Heat Transfer Through A Turbulent Boundary Layer. International Journal of Heat and Mass Transfer 1991; 34, 1917-1928.

4. Harris S, Ingham D, Pop I. Transient Boundary Layer Heat Transfer From A Flat Plate Subjected To A Sudden Change in Heat Flux. European Journal of Mechanics B- Fluid 2001; 20, 187-204.

5. Deswita L, Nazar R, Ahmad R, Ishak A, Pop I. Similarity Solutions of Free Convection Boundary Layer Flow on a Horizontal Plate with Variable Wall Temperature. European Journal of Scientific Research 2009; 27(2) 188-198.

6. Defraeyea T, Houvenaghelc G, Carmelieta J, Derome D. Numerical Analysis of Convective Drying of Gypsum Boards. International Journal of Heat and Mass Transfer 2012; 55, 4487-4928.
7. Mori S, Nakagwa H, Tanimoto A, Sakakibara M. Heat and Mass Transfer With A Boundary Layer Flow Past A Flat Plate of Finite Thickness. International Journal of Heat and Mass Transfer 1991; 34, 2899-2909.
8. Masmoudi W, Prat M. Heat and Mass Transfer Between A Porous Medium and A Parallel External Flow; Application to Drying of Capillary Porous Material. International Journal of Heat and Mass Transfer 1991; 34, 1975-1989.
9. Jomaa W, Bruneau D, Nadeau J. Simulation of The High Temperature Drying of A Past Product: on The Influence of The Local Air Flow and The Thermal Radiation. Drying Technology 2004; 22, 1709-1729.
10. Ranz W, Marshall W. Evaporation From Drops. Chemical Engineering Progress 1952; 48, 141-173.
11. Trommelen A, Crosby E. Evaporation and Drying of Drops in Superheated Vapors. AIChE Journal 1970; 16, 857-872.
12. Sirignano, W.A. Fluid Dynamics and Transport of Droplets and Sprays: Cambridge University Press, NY, USA 1999.
13. Masters K. Spray Drying in Practice. Spray Dry Consult International: Denmark: John Wiley & Sons Inc., NY, USA 2002.
14. Sloth J, Kiil S, Jensen A, Andersen S, Jørgensen K, Schiffter H, Lee G. Model Based Analysis of The Drying of A Single Solution Droplet in An ultrasonic Levitator. Chemical Engineering Science 2006; 61, 2701-2709.
15. Miura K, Miura T, Ohtani S. Heat and Mass Transfer to and From Droplets. American Journal of Chemical Engineers, Symposium series 1977; 73 (163).
16. Ali H, Mumford C, Jefferys G., Bains G. A study of evaporation from, and drying of, single droplets. In: Mujumdar AS. (editor) Proceedings of the 6th International Symposium in Drying IDS'88. Versailles, France.1988.

17. Minoshima H, Matsushima K, Liang H, Shinohara K. Estimation of Diameter of Granule Prepared by Spray Drying of Slurry with Fast and Easy Evaporation. Journal of Chemical Engineering of Japan 2002; 35, 880-885.

18. Seydel P. Experimental and Mathematical Modeling of Solid Formation at Spray Drying. Chemical Engineering Technology 2004; 27 (5): 505-510.

19. Perdana J, Fox M, Schutyser M, Boom R. Single-Droplet Experimentation on Spray Drying: Evaporation of a Sessile Droplet. Chemical Engineering & Technology 2011; 34 (7): 1151–1158.

20. Cheong H, Jeffreys G, Mumford C. A Receding Interface Model For The Drying of Slurry Droplets. AIChE Journal 1986; 32, 1334-1346.

21. Farid M. A New Approach to Modeling of Single Droplet Drying. Chemical Engineering Science 2003; 58, 2985-2993.

22. Almubarak A, Mumford C. Characteristics of the receding evaporation front in convective drying. In: Mujumdar AS. (editor) Proceedings of the 9th International Drying Symposium, IDS'94, Marcel Dekker Inc., NY, USA; 1994.

23. Kays W. M, Crawford M.E. Convective Heat and Mass Transfer, 2nd. ed., McGraw-Hill Book Company, New York, NY, USA.1980.

24. Thomas L. C. Heat Transfer professional Version. 2nd ed.: Capston Publishing Corporation, OK. USA. 1999.

25. Kreith F, Bohn M. Principles of Heat Transfer. 6th ed.: Brook/Cole Publishers, Pacific Grove, CA, USA 2001.

26. Almubarak A, Al-Saeedi J, Shoukry M. Effect of Boundary Layer on Mechanisms of Beach and Desert sand. European Journal of Soil Science 2008; 59, 807-816.

27. Waananen K, Litchfield J, Okos M. Classification of Drying Models for Porous Solids, Drying Technology 1993; 11, 1-40.

28. Holman J.P. Heat Transfer: McGraw-Hill Book Co., NY, USA 2002.

29. Nesic S, Vodnik J. Kinetics of Droplet Evaporation. Chemical Engineering Science 1991; 46, 527-537.

30. Chen X, Peng F. Modified Biot Number in The Context of Air Drying of Small Moist Porous Objects. Drying Technology 2005; 23, 83-103.

31. Hayder M, Mumford C. 1993. Mechanisms of Drying of Skin Forming Materials. Drying Technology 1993; 11, 1713-1750.

9

CHAPTER

Pose Measurement Method and Experiments for High-Speed Rolling Targets in a Wind Tunnel

Zhenyuan Jia [1], Xin Ma [1], Wei Liu [1,*], Wenbo Lu [2], Xiao Li [1], Ling Chen [1], Zhengqu Wang [2] and Xiaochun Cui [2]

[1]*Key Laboratory for Precision and Non-Traditional Machining Technology of the Ministry of Education, Dalian University of Technology, Dalian 116024, China*

[2]*AVIC Aerodynamics Research Institute, Shenyang 116024, China External Editor: Vittorio M.N. Passaro*

ABSTRACT

High-precision wind tunnel simulation tests play an important role in aircraft design and manufacture. In this study, a high-speed pose vision measurement method is proposed for high-speed and rolling targets in a supersonic wind tunnel. To obtain images with high signal-to-noise ratio and avoid impacts on the aerodynamic shape of the rolling targets, a high-speed image acquisition method based on ultrathin retro-reflection markers is presented. Since markers are small-sized and some of them may be lost when the target is rolling, a novel markers layout with which markers are distributed evenly on the surface is proposed based on a spatial coding method to achieve highly accurate pose information. Additionally, a pose acquisition is carried out according to the mentioned markers layout after removing mismatching points by Case Deletion Diagnostics. Finally, experiments on measuring the pose parameters of high-speed targets in the laboratory and in a supersonic wind tunnel are conducted to verify the feasibility and effectiveness of the proposed method. Experimental results indicate that the position measurement precision is less than 0.16 mm, the pitching and yaw angle precision less than 0.132° and the roll angle precision 0.712°.

Keywords: machine vision; pose measurement; high-speed motion; wind tunnel

1. INTRODUCTION

The pose measurement method has been widely used in many research fields such as aerospace, robots and medical science. In some cases, the improper separation of stores carried by a high-speed aircraft, such as bombs, auxiliary fuel tank, torpedoes and missiles, will greatly impact the performance of the mother plane and even threaten the safety of the pilot [1,2]. Simulative store separation experiments in wind tunnels provide valuable reference for the layout design and release parameters of the store. However, since wind tunnel models are usually small-sized and move and rotate at high speeds, traditional measurements fail to correctly measure their pose information. Vision measurement has proved its advantages such as being non-contact and high precision and the ability for real-time measurement. Therefore, pose measurement in wind tunnels based on vision measurement is of great importance [3,4].

Numerous approaches for measuring the pose parameters of targets have been studied by many scholars and research institutions. Watzlavick *et al.* presented the Optotrak™ measurement system by Canada's Northern Digital Inc. (Waterloo, ON, Canada). Three precisely calibrated linear CCDs are used as displacement sensors [5]. The 3D coordinates of each marker can be obtained in real time by capturing the near-infrared light given out by the active luminous markers. The precision and resolution of the system can reach 0.1 mm RMS and 0.01 mm, respectively. It has been successfully applied to the aerodynamic elasticity and angle measurement of the wind tunnel models of both Boeing and the NASA Ames Research Center. However, its low sampling frequency and big volume of embedded markers cannot satisfy high-speed pose measurement demand of small volume objects. Rock *et al.* combined the Carrier-Phase Differential Global Positioning System (CDGPS) and a binocular vision system to identify ground target images [6]. They used color object identification technology to process images captured by the left and right cameras, and employed two different images to locate the target projection. Finally, the aircraft pose parameters can be estimated according to stereo triangulation. Though this method is simple in calculation, the wireless signals in a complex environment are susceptible to be disturbed, so the measurement data are not stable. NASA LaRC developed a pose measurement system based on binocular vision for the wind tunnel model [7–10]. The model images are captured continuously and synchronously by two CCD cameras through the observation window. The pose parameters of the wind tunnel model can be calculated after measuring the 3D centroid coordinates of circular markers. However, this method has disadvantages due to the hidden phenomenon when the model is rolling. Murray *et al.* employed an Optical Tracking System based on Particle Image Velocimetry (PIV) to obtain the pitch angle, angular velocity and related pose information of a high-speed target [11]. Though high-speed measurements can be achieved, they cannot get the rolling angle utilizing this system. Chen presented a pose measurement method for experimental model based on active vision in a wind tunnel [12]. This method transforms the direct pose measurement of the target into an indirect measurement of spots on the wall, which amplifies the variation of the information. Thus, the resolution of the measurement system can be indirectly improved, and the maximum error of displacement of the cooperating target is 0.46 mm. However, its low precision cannot satisfy pose measurement requirements of small volume high-speed target and high-speed rolling motion. Fan *et al.* proposed an attitude and position tracking system for indoor carriers based on an integrated navigation fusion of INS/UWB information strategy [13]. The two subsystems (INS and UWB) are integrated by a feedback correction method. Then, the data fusion of the two subsystems can be realized by an optimal comprehension and filtering strategy. Though the position and attitude information can be obtained in real time, when it comes to small targets with high-speed rolling motion in a wind tunnel, the UWB technology can hardly meet the measuring environment demands. In summary, scholars have made different kinds of investigations on innovations in pose measurement under wind tunnel conditions, but the pose measurement speed is low and the hidden phenomenon of cooperation markers in the rolling procedure is ignored.

In this paper, a pose measurement method for high-speed rolling targets in wind tunnels is presented. The remainder of this paper is organized as follows: Section 2 will detail the vision theory and image acquisition method; Section 3 will describe the marker layout on the surface of target and the image matching method; Section 4 will present the pose information acquisition method from the constructed markers' 3D coordinates; Section 5 will show the results obtained from a set of testing scenarios. Finally, the paper ends with conclusions about the whole pose measurement method and the results obtained.

2. MEASUREMENT METHOD AND SYSTEM FOR HIGH-SPEED AND ROLLING TARGETS

2.1. Binocular Vision Theory

Vision measurement focuses on how to obtain 3D information from 2D image information. Two types of coordinate systems (*i.e.*, image and global coordinate systems) are involved in transforming the 2D image coordinates obtained from the camera into 3D global coordinate information, as shown in Figure 1.

In this study, the pinhole model is taken as the imaging model. However, due to some defects occurred in the manufacture and installation procedures of composite lenses, image distortion is unavoidable, resulting in aberration between the real and theoretical image points. Therefore, we introduce a non-linear imaging model considering the impact of aberration. The projection equation of the non-linear imaging model can be given by:

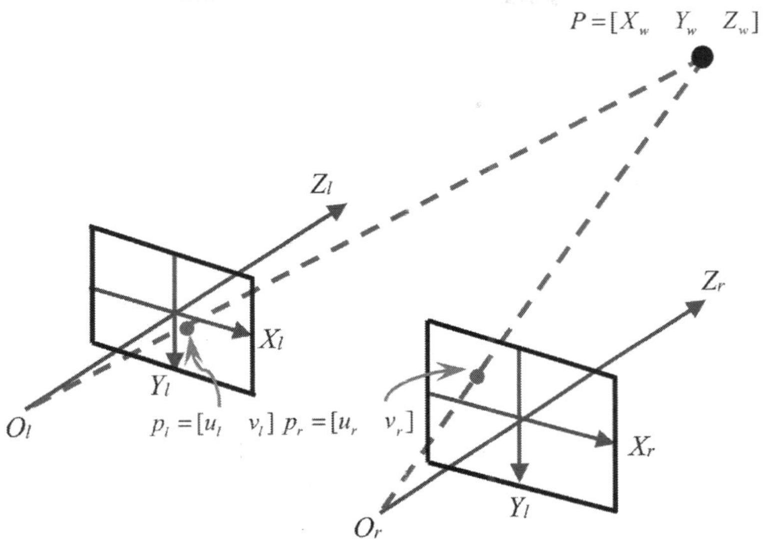

Figure 1. The principle of binocular stereo imaging.

$$Z_c\begin{bmatrix}\hat{u}+\delta_x\\ \hat{v}+\delta_y\\ 1\end{bmatrix}=M_1\begin{bmatrix}R & T\\ 0^T & 1\end{bmatrix}\begin{bmatrix}X_w\\ Y_w\\ Z_w\\ 1\end{bmatrix} \quad (1)$$

where $p(\hat{u}\ \hat{v})$ represents the real point caused by image distortion; $(\delta_x\delta_y)$ stand for aberration, which are known already; $(X_wY_wZ_w)$ represent the 3D coordinates of one point in the world coordinates; Z_c is the scale factor; M_1 represents the intrinsic parameter of the camera; rotation matrix R and translation matrix T are the extrinsic parameters.

Wide-angle lenses are used in our measurement system to obtain close shots with a large field of view, thus the main image distortion in our system is barrel distortion. To obtain a computation method with less complex and better stability, the first and second order distortion coefficients are employed in the aberration model, as shown below:

$$\begin{cases} \delta_x = k_1 \hat{u}\left(\hat{u}^2 + \hat{v}^2\right) + k_2 \hat{u}(\hat{u}^2 + \hat{v}^2)^2 \\ \delta_y = k_1 \hat{v}\left(\hat{u}^2 + \hat{v}^2\right) + k_2 \hat{v}(\hat{u}^2 + \hat{v}^2)^2 \end{cases} \tag{2}$$

where k_1, k_2 are the first and second coefficient of radial distortion.

Camera calibration is the preliminary step for establishing an accurate relationship between the image and the real world. In this paper, Zhang's 2D calibration method is implemented to achieve a high precision calibration of the binocular vision measurement system with a wide field of 1 m × 1 m [14]. The calibration targets are shown in Figure 2.

Figure 2. Calibration targets taken in different angles.

2.2. Measurement Method

In this paper, position and attitude parameters of high-speed and rolling targets can be achieved *via* analyzing image sequences of markers' moving trajectory which is captured in real time by a binocular vision measurement system. The measurement principle for position and attitude parameters of high-speed moving targets is depicted in Figure 3. Firstly, ultrathin retro-reflection markers are attached on the surface of target. Then, the left and right CCD cameras are used to capture the image sequences by applying the high-speed image acquisition technique. Moreover, the high-precision matching and recognition of the corresponding markers in the left and right images are facilitated based on spatial encoding, and then the 3D coordinates of the markers in the world coordinate system are reconstructed. Furthermore, the axis of the slender rolling target is fitted using Case Deletion Diagnostics. Finally, the pose information of the target is obtained by computing the 3D coordinates of the markers.

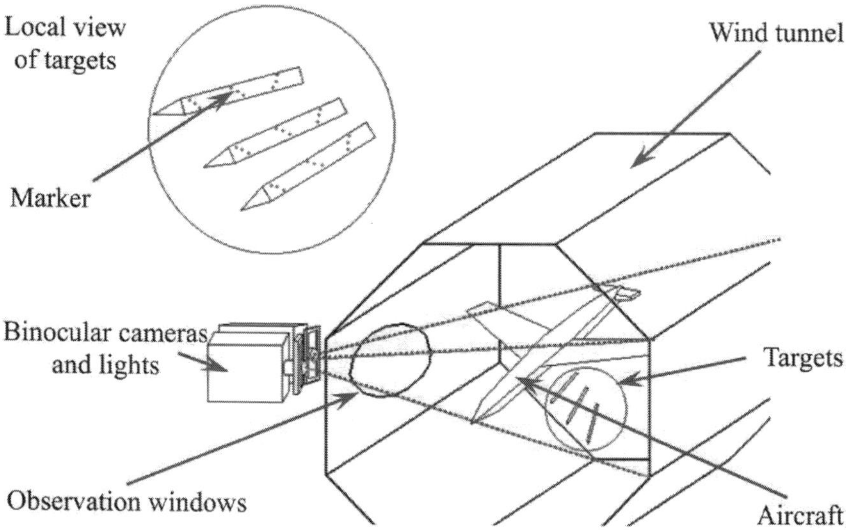

Figure 3. Camera-tunnel arrangement.

2.3. Ultrathin Retro-Reflection Markers for High-Speed Image Acquisition

The high speed image acquisition of rolling targets in a wind tunnel mainly has the following difficulties: (a) Owing to the limited size of the observation window and the high image acquisition speed, it is difficult to guarantee clear images; (b) Since the target is mostly a small-sized cylinder-like object of complicated shape, it is difficult to ensure the position accuracy of markers; (c) In order to guarantee the aerodynamic shape of the scaled model, the thickness of markers cannot exceed 30 um. However, normal markers are too thick (160 μm) to meet the requirements. To solve the aforementioned problems, an ultrathin retro-reflection markers fabrication technique based on a rolling-printed method is proposed to allow the high speed image acquisition of high-speed rolling target. Ultrathin retro-reflection markers are made by mixing glass microbeads, silver powder and resin according to the specific mixing ratio of 16:5:9. The thickness of the resulting ultrathin retro-reflection markers is only 20 μm, which guarantees the aerodynamic shape of the scaled model.

The diagram of the rolling-printed method is shown in Figure 4, where the screen plate that is designed according to the marker layout is placed atop the target. The ultrathin retro-reflection material is placed on top of the screen plate. When a process performs an operation on the device, the screen plate keeps in pure rolling contact with the target and moves continuously at a certain linear speed V along the peripheral velocity direction. V can be computed by $V = \omega r$, where ω and r depict the angular velocity and radius of the rolling target, respectively. As a result, the ultrathin retro-reflection material in the mesh opening can be pumped or squeezed by the squeegee on the surface of the target. Figure 5 presents the finished target.

In this way, the thickness and the positioning accuracy of markers are ensured. Besides, images captured by the left and right cameras have higher signal-to-noise ratio and better brightness, and the image acquisition requirements can be satisfied under high shooting frequency conditions. As shown in Figure 6, the brightness of markers decreases while the shutter speed increases. However, markers are brighter and have higher signal-to-noise ratio at the 5000 shooting speed.

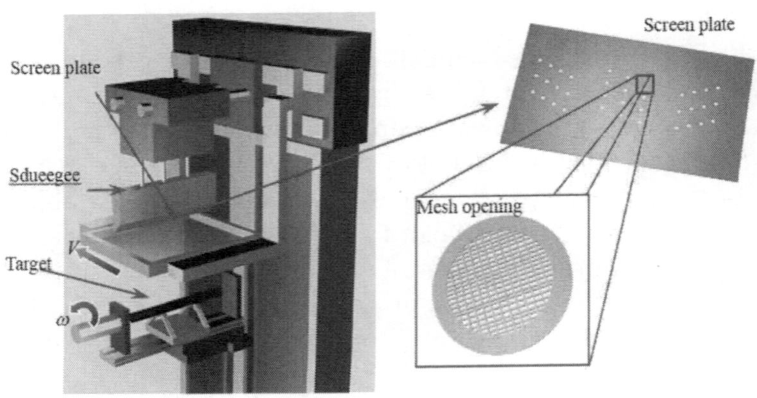

Figure 4. Diagram of the rolling-printed method and structure of the screen plate.

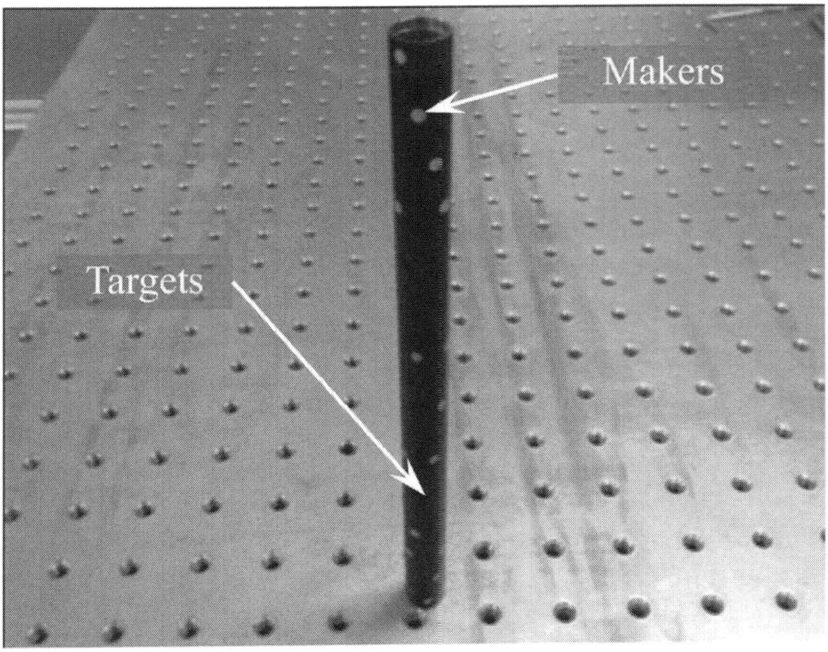

Figure 5. The finished target.

Figure 6. Shooting picture of ultrathin retro-reflection markers in different frames.

3. IMAGE PROCESSING

In this paper, cooperating markers are adopted to achieve rapid pose calculation of rolling targets. In order to improve the stability and precision of the results, it is necessary to attach more markers on the surface of the target. However, due to the small imaging area and lack of feature information, it is hard to match and recognize these markers. This section describes how to match and recognize markers with information loss.

3.1. Markers Layout of High-Speed Rolling Target

In order to calculate the pose parameters of the moving target quickly and accurately, the following problems must be solved: (1) it is difficult to acquire accurate matching and recognition of markers with less imaging area and inconspicuous motion features; (2) the high-speed rolling target in this paper is long and thin, making the target surface really small. Consequently, marker distribution is relatively dense, so that the recognition and matching of large number of markers would be very difficult; (3) markers' features (quantity, serial number, layout, *etc.*) are also difficult to be ascertained during the measuring process owing to the rolling motion of the target. To solve those problems, a spatial encoding method is presented in this paper.

Traditional encoding methods complete the encoding process by changing the shape of a single marker. Combining with dividing several markers into one group based on spatial relation, positional relationship between markers in the same group is used in the spatial encoding method proposed in this paper to realize encoding process [15]. As a result, the problem that small markers in larger field of view are difficult to encode is solved. As shown in Figure 7, each three markers are divided into one group which is called a spatial encoding group according to a linear constraint, and then the three markers can be ascertained through the distance ratio between them. The formula can be given by:

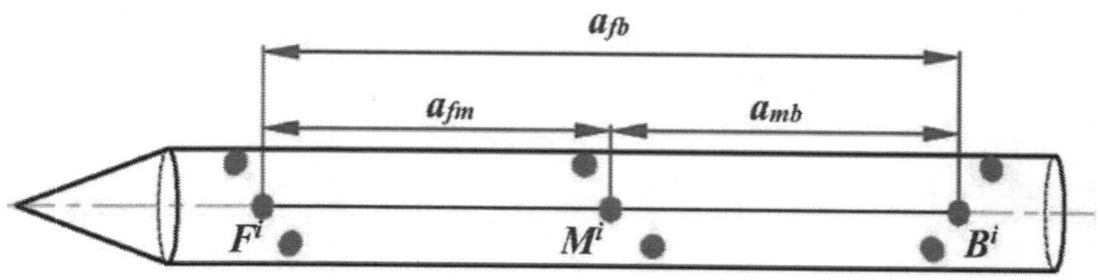

Figure 7. Markers distributed along the circumferential direction of the target.

$$\frac{a^i_{fm}}{a^i_{mb}} = c_i, \frac{a^i_{fm}}{a^i_{mb}} \neq \frac{a^j_{fm}}{a^j_{mb}} \tag{3}$$

$$\frac{a^i_{fm}}{a^i_{fb}} = l_i, \frac{a^i_{mb}}{a^i_{fb}} = n_i, \frac{a^i_{fm}}{a^i_{fb}} \neq \frac{a^i_{mb}}{a^i_{fb}} \tag{4}$$

where a^i represents distance between the two space encoding markers in the same group of i; f, m and d stand for head, middle, tail part of the target, respectively; let a^i_{fm} be the distance between two markers in the head and middle part of the target; l_i is the distance ratio between a^i_{fm} and a^i_{fb}, which can distinguish the ith group from all other groups; c_i is the distance ratio between a^i_{fm} and a^i_{mb}; n_i is the distance ratio between a^i_{mb} and a^i_{fb}; in this paper, we defined V be encoding array, then the encoding array of group i can be written as $V_i = [l_i c_i n_i]$, and the spatial encoding of markers can be accomplished by making the array elements in every encoding array unique.

To improve the precision and stability of the measurement system, it is necessary to increase the number of markers in the layout to satisfy the extraction precision. Moreover, a radial circular uniform distribution is adopted to solve the problem of the rolling motion of the target. To meet the requirements of image acquisition, a total of 12 spatial encoding groups are attached onto the target accurately according to the distance information of these three markers, as shown in Figure 8.

In the process of measuring the pose of the target, at least three groups with a total of nine markers appear in the image simultaneously. The well-proportioned distribution of markers in the head, middle, and tail part of the target guarantees the number of markers as well as the pose estimation precision of the target.

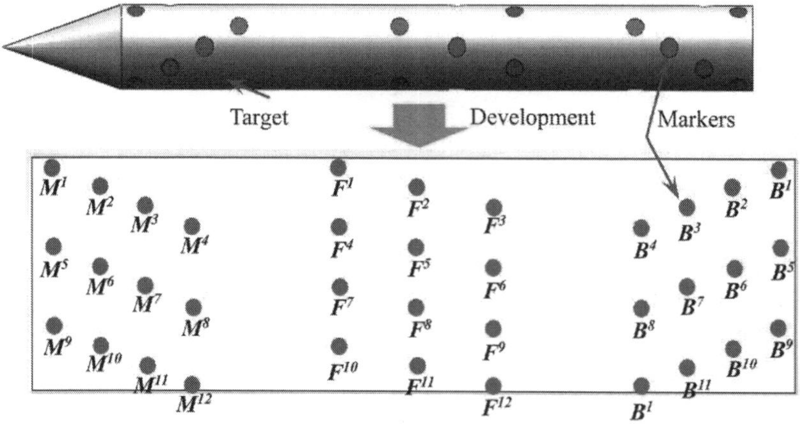

Figure 8. The markers layout.

3.2. Feature Extraction

In the complicated wind tunnel conditions, it is difficult to conduct the marker feature extraction due to such factors as small-sized model with large curvature change, large field of view and small imaging area of markers. Therefore, the gradient weighted method is employed to extract the center of the markers with high accuracy. Firstly, a convolution operation of the image is conducted based on a Gaussian first-order differential operator in order to obtain the gradient of each point in the image. The Gaussian first-order differential operator can be written as:

$$h'(x) = \frac{-x}{\sqrt{2\pi\sigma}} e^{-\frac{x^2}{2\sigma^2}} \tag{5}$$

Then, the gradient of image can be given by:

$$|G(x,y)| = \begin{bmatrix} F_x' \\ F_y' \end{bmatrix} = \begin{bmatrix} F(x,y) \otimes h_x' \\ F(x,y) \otimes h_y' \end{bmatrix} \tag{6}$$

The coordinates of gradient centroid of markers pattern can be expressed as:

$$c=\sum_{i=-h}^{h}\sum_{j=-w}^{w}\left[\left|G(i,j)\right|\cdot P(i,j)/\sum_{i=-h}^{h}\sum_{j=-w}^{w}G(i,j)\right] \quad (7)$$

where c is the coordinates of the gradient centroid of the markers; $|G(i,j)|$ represents the weight of point (i,j); w and h stand for the width and height of target image, respectively; $P(i,j)$ is the image coordinate of the point (i,j).

3.3. Feature Matching between Two Views

Gray matching and feature matching are generally used for the classical and effective image matching. The gray matching method is a mathematical procedure based on the gray information and image technique, while features such as color, texture, shape, spatial position extracted from two or more images are used in feature matching. However, high-precision matching under extreme environment conditions cannot be achieved according to the gray and feature information due to the fact that markers captured by ultra high-speed cameras have a small area and the gray similarity between each one is big.

To solve the problem, spatial encoding markers are applied to transform the traditional matching method based on features into a problem according to spatial constraints between markers, as shown in Figure 9.

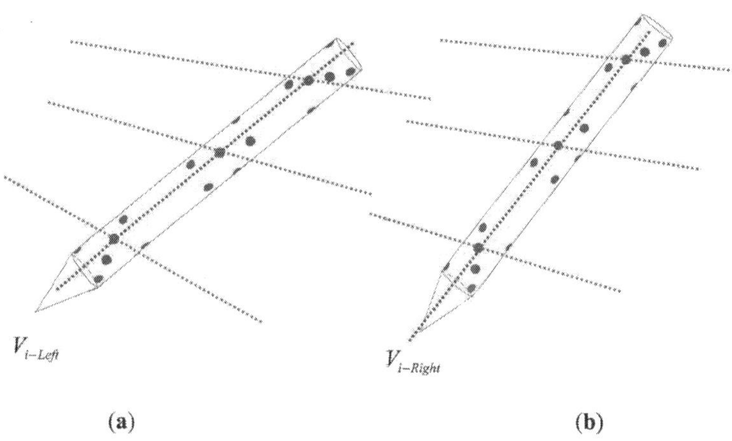

Figure 9. Matching between left and right images.

Firstly, spatial encoding markers in different groups are isolated by a collinear constraint, and then the matching relationship between the left and right images can be obtained through comparing $V_{i\text{-}Left}$ (obtained in Section 3.1) of the right image and $V_{i\text{-}Right}$ of left image. Moreover, based on the epipolar constraints, the corresponding marker matching is verified and the mismatched markers are removed.

3.4. The Recognition Method

The recognition process can be described as follows: (a) Reconstructed markers are grouped according to Section 3.3; (b) Recognition relationships can be obtained by comparing in turn each encoding array V obtained by construction (Section 3.1) with the one V_i (i=1,2...12) determined by markers layout, as shown in Figure 10.

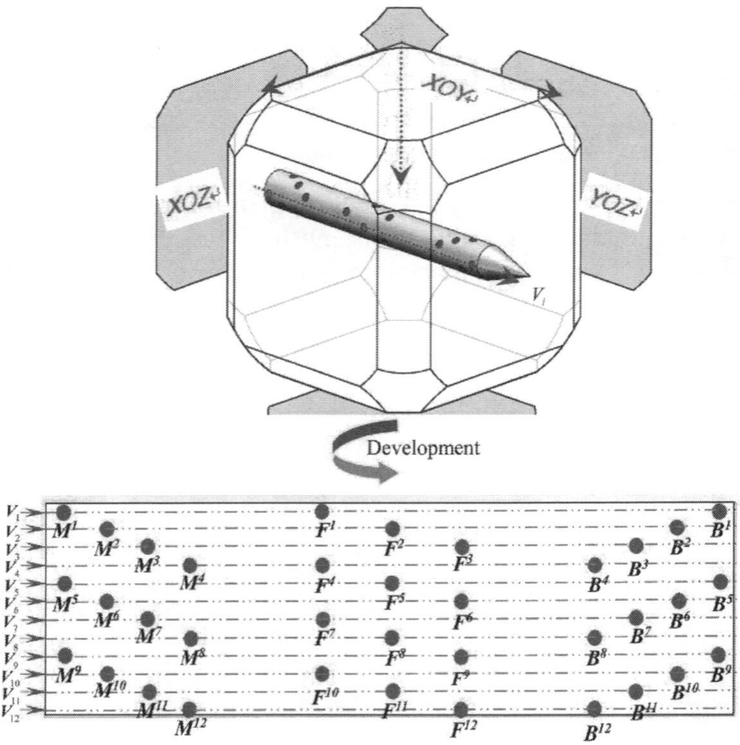

Figure 10. Spatial model of marker recognition.

4. POSITION AND ATTITUDE SOLUTION

The position and attitude parameters can be obtained by translating the coordinates from the local coordinate system on the slender rolling target to the world coordinate system through the translation and rotation matrix. Firstly, the axis of the slender rolling target is fitted by Case Deletion Diagnostics. Then the coordinates of the centroid can be solved based on the distance and axis constraint. Moreover, the absolute position method is applied to calculate the initial translation and rotation coordinate system which are used for the transformation, and at last the position and attitude information can be obtained.

4.1. Axis Fitting Based on Case Deletion Diagnostics

Errors caused by mismatching points have a great influence on the axis fitting solution. Thus, to guarantee the fitting precision, an axis fitting method based on Case Deletion Diagnostics is proposed to remove mismatched points. Case Deletion Diagnostic is commonly used to assess the influence of observations on the parameter estimators through subset deletion [16]. Utilizing the Case Deletion Diagnostic, the outlier or strong influence points are diagnosed by comparing the statistic difference between the deleted model and un-deleted model, thus the gross errors can be removed. Specifically, suppose that the data is $P=\{P_1,P_2,...,P_n\}$, when removing the *i*th data point, if there is a difference between the deleted model and the un-deleted model, the outlier or strong influence points can be obtained, otherwise we can acquire the correct points which have little impact on statistical results.

Let the total number of the markers be *n* and at least nine markers can be collected according to markers layout. Then the point set under the world coordinate system can be given by:

$$P=\{P_1,P_2,\cdots,P_n\} \tag{8}$$

$$P_i=(x_i,y_i,z_i),(i=1,2,\cdots,n) \tag{9}$$

where (x_i, y_i, z_i) describe the 3D coordinates of point P_i in the world coordinate system. P stands for a point set. Case Deletion Diagnostic $M_w(\theta)$ for axis fitting can be set up as follows:

1) The point set is applied to fit the axis equation L according to the theoretical radius r which represents the distance between markers and axis.

2) he parameter estimators θ can be expressed as:

$$\theta = \sqrt{\sum_{i=1}^{n} \left(d\left(p_i\right) - r\right)^2} \tag{10}$$

where $d(p_i)$ is the euclidean distance from point p_i to axis L; axis $\hat{L}$ for slender rolling target can be obtained after deleting the point p_w.

3) The estimated value $\hat{\theta}$ of parameter estimators can be given by:

$$\hat{\theta} = \sqrt{\sum_{i=1, i \neq w}^{n} \left(\hat{d}\left(p_i\right) - r\right)^2} \tag{11}$$

The judgment that whether the point p_w is outlier or strong influence point can be expressed as:

$$\left|\hat{\theta} - \theta\right| \leq \varepsilon \tag{12}$$

where ε represents the threshold defined above. When the above inequality is satisfied, we can get the result that point p_w is a normal point, otherwise it is a strong influence point. The initial point set for Case Deletion Diagnostic $M_w(\theta)$ is all the markers collected. Every point is judged one by one, and a point within the set can be removed if it is outlier or strong influence point. Afterwards, the axis can be solved stably using the LM optimizing algorithm:

$$\begin{cases} \min \sqrt{\sum_{i=1}^{n} \left(d(p_i) - r\right)^2} \\ s.t. \quad \dfrac{x - x_0}{m} = \dfrac{y - y_0}{n} = \dfrac{z - z_0}{p} \\ d(p_i) = \dfrac{\sqrt{\begin{vmatrix} y_i - y_0 & z_i - z_0 \\ n & p \end{vmatrix} + \begin{vmatrix} z_i - z_0 & z_i - x_0 \\ p & m \end{vmatrix} + \begin{vmatrix} x_i - x_0 & y_i - y_0 \\ m & n \end{vmatrix}}}{\sqrt{m^2 + n^2 + p^2}} \end{cases} \tag{13}$$

where $P_0(x_0,y_0,z_0)$ stands for one point on the fitted axis; $L(m,n,p)$ is the direction vector of axis; $d(p_i)$ represents the distance between point $P_i(x_i,y_i,z_i)$ and the fitted axis.

4.2. Solution for Centroid Position Based on Distance Constraint

The centroid coordinates can be solved based on the distance constraint between markers. In this paper, about nine to twelve markers can be extracted according to the layout design. In order to improve the centroid coordinates' precision, optimization based on the distance between all markers captured and centroid are carried out. However, the aforementioned distance is different, and the extracted markers have different degrees of brightness and circularity. These factors have a different impact on the precision of the centroid coordinates. Thus, multi-points optimization based on setting weight for every point is presented. Since the centroid is on the axis, the centroid position can be optimized according to axis position calculated *via* Section 4.1. The optimization formula can be given by:

$$\begin{cases} \min \left(\sum_{i=1}^{n} \rho_i (Dc_i - dc_i)^2 \right) \\ s.t. \quad \dfrac{x_c - x_0}{m} = \dfrac{y_c - y_0}{n} = \dfrac{z_c - z_0}{p} \end{cases} \tag{14}$$

where ρ_i represents the weight value of each captured marker; Dc_i stands for the measured distance between the centroid and the optimized markers; dc_i is the real distance between the centroid and optimized markers; $C(x_c,y_c,z_c)$ is the centroid coordinate; $L(m,n,p)$ stands for the direction vector of the axis.

4.3. Absolute Pose Solution

In this paper, a pose acquisition method is carried out according to the aforementioned marker layout. Two coordinate systems–the initial reference coordinate system (IRCS) and the local coordinate system (LRCS), are established to acquire the pose information of the target through coordinate transformation. The IRCS represents the reference coordinate system set up at the first frame, while the LRCS means the local coordinate system set up at the frame of the measuring position.

As shown in Figure 11, when the axis fitting is finished (Section 4.1) and the centroid position determined (Section 4.2), the centroid of target P_1 is set as the origin of the reference coordinate system and the axis of the target is defined as the Y axis of the reference coordinate system.

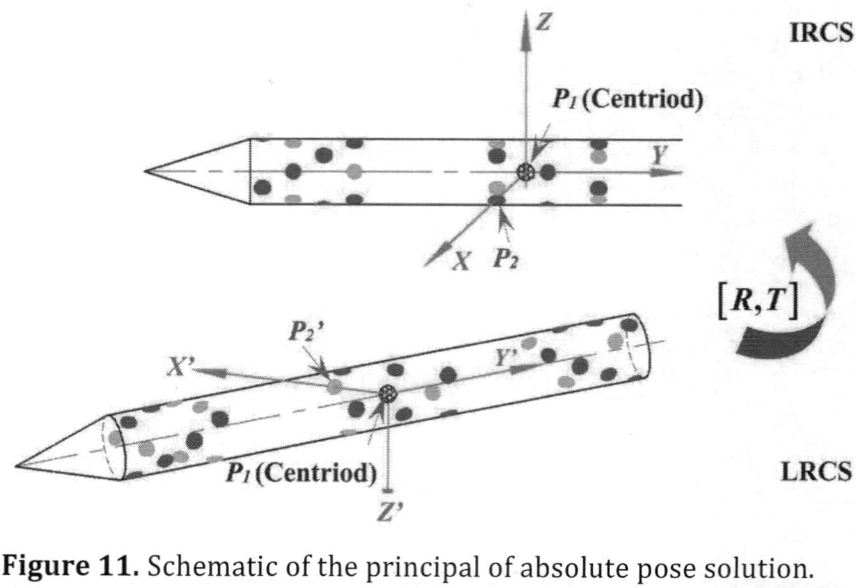

Figure 11. Schematic of the principal of absolute pose solution.

Moreover, since the X axis can be defined by any marker captured by cameras with high image quality, the optimum marker P_2 is chosen as the definition point. As a result, the space line through the centroid of the target parallel to the line through P_2 and perpendicular to the rotation axis of the target is defined as the X axis. Then, according to the right-handed coordinate system, the reference coordinate system can be set up.

To obtain the pose parameters, it is necessary to establish local coordinate system using the same marker P_2. However, marker P_2 may not be captured at the position measuring frame owing to the hidden markers phenomenon caused by the rolling movement. In order to establish the local coordinate system, the three-dimensional coordinates of all visible makers at the frame of the measuring position are reconstructed. Then, based on the recognition of visible markers and markers layout, the three-dimensional coordinates of the center of marker P_2 can be acquired. Afterwards, the local coordinate system can be obtained using the same establishment principle for the reference coordinate system.

The relationship between the point $p = (x\ y\ z)^T$ in IRCS and the corresponding $p' = (x'y'z')^T$ in LRCS can be written by:

$$\begin{bmatrix} x' \\ y' \\ z' \end{bmatrix} = R_{pt} \begin{bmatrix} x - x_0 \\ y - y_0 \\ z - z_0 \end{bmatrix} \tag{15}$$

where $(x_0\ y_0\ z_0)$ is the coordinates of the origin of the target in IRCS. Moreover:

$$R \begin{bmatrix} x \\ y \\ z \end{bmatrix} + T = \begin{bmatrix} x' \\ y' \\ z' \end{bmatrix} \tag{16}$$

where:

$$R = R_{\theta_Y} R_{\theta_X} R_{\theta_Z} = \begin{bmatrix} \cos(\theta_Y)\cos(\theta_Z) - \sin(\theta_Y)\sin(\theta_X)\sin(\theta_Z) & \cos(\theta_Y)\sin(\theta_Z) + \sin(\theta_Y)\sin(\theta_X)\cos(\theta_Z) & -\sin(\theta_Y)\cos(\theta_X) \\ -\cos(\theta_X)\sin(\theta_Z) & \cos(\theta_X)\cos(\theta_Z) & \sin(\theta_X) \\ \sin(\theta_Y)\cos(\theta_Z) + \cos(\theta_Y)\sin(\theta_X)\sin(\theta_Z) & \sin(\theta_Y)\sin(\theta_Z) - \cos(\theta_Y)\sin(\theta_X)\cos(\theta_Z) & \cos(\theta_Y)\cos(\theta_X) \end{bmatrix} \tag{17}$$

As a result, the centroid position of the target in the world coordinate system can be represented by T_{pb}= $(x_0\ y_0\ z_0)$; $-\theta_Z$, $-\theta_X$, $-\theta_Y$ represent the three attitude parameters of the target in the world coordinate system: the yaw, pitch and roll. According to the coordinates of the chosen points mentioned above, the position and attitude parameters of the target can be computed.

5. EXPERIMENTAL ANALYSIS

Experiments on pose measurement of the target in a wind tunnel are conducted in this paper, as well as the precision verification test. In this section, the measurement precision of the three directions and three angles is verified.

5.1. Vision Measurement System

A wind tunnel is a dark narrow enclosure which can only be monitored by an observation window with a fixed shape. According to the actual wind tunnel structures, an ultra high-speed binocular vision measurement system is set up, as shown in Figure 12. The binocular vision measurement system contains two high-speed cameras (FASTCAM SA-X), two wide-angle lenses (Nikon 1735), two low angle lights, a shockproof platform, two electronic control platforms and a processing terminal. To minimize the influence of vibrations caused by the wind tunnel at runtime, the two cameras are placed symmetrically on a customized air flotation shockproof table. The camera system is installed on the electric control platform so that it can be translated and rotated as designed. The low angle DC lights are installed on the edge of the lenses to ensure the light requirements of images are met.

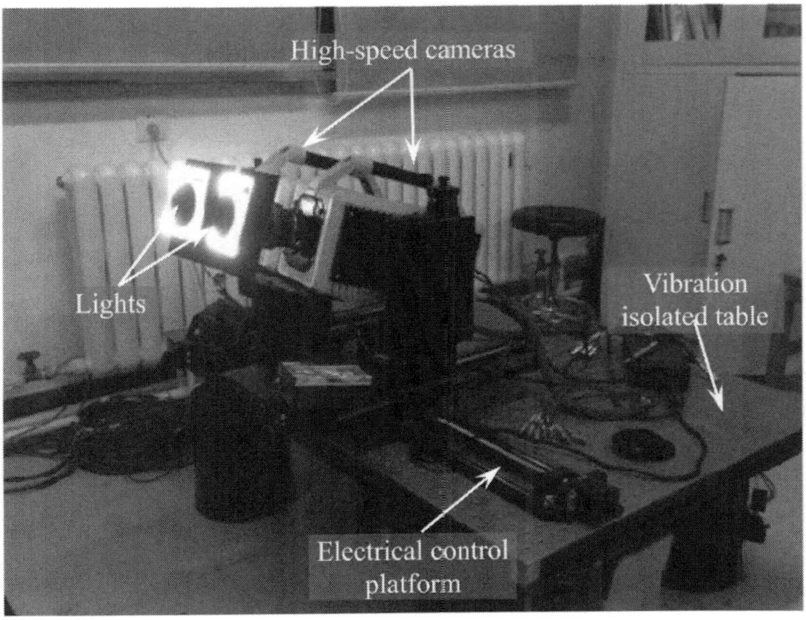

Figure 12. Mechanical support and hardware for cameras.

5.2. Measurement Experiment for Rotary Target in Wind Tunnel

As shown in Figure 13, the experimental facilities consist of a pose measurement system based on binocular vision, an ejection mechanism and a controlling system.

In the test with measurement field of 1 m × 1 m and measurement rate of 5000 fps, the ejection mechanism and controlling system are utilized to eject the target out of the tank with a particular velocity, angular velocity and angle. The vision measurement system is triggered to measure the pose information of the target synchronously. Based on the aforementioned method, the pose measurement resulte of the rolling target are verified.

To calculate the actual pose parameters of the target, experiments were conducted in a closed supersonic wind tunnel (FL-3). The facility layout for the wind tunnel tests is shown in Figure 14.

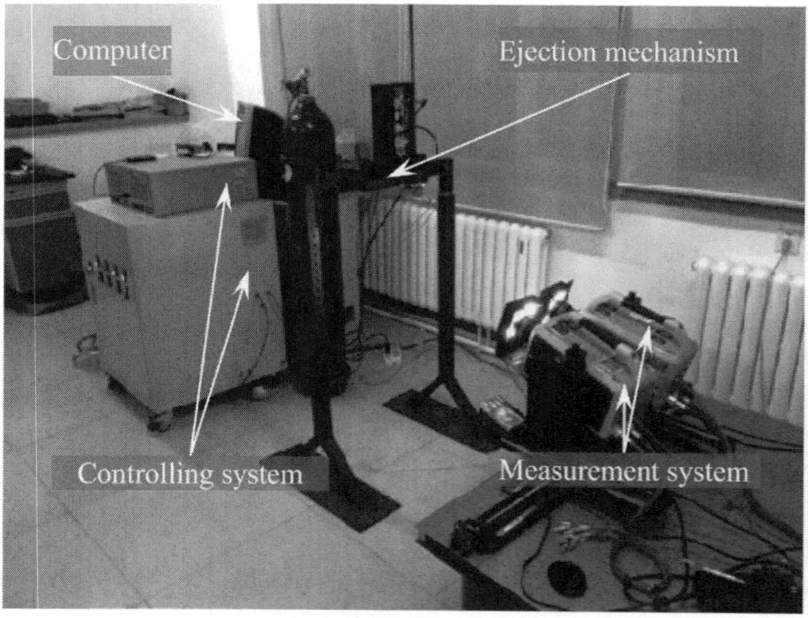

Figure 13. Layout of experimental facilities.

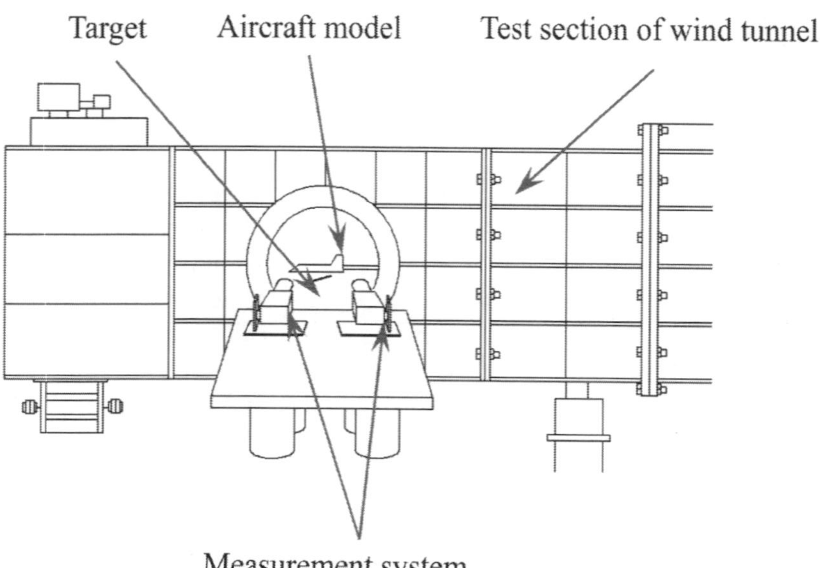

Figure 14. Facility layout for the wind tunnel test.

Specifically, an aircraft model is installed in the inner wall of the test section, and the ejection mechanism is installed in the ejection cabin of the aircraft model. Besides, the measurement system is arranged outside of the test section to measure the pose parameters of the object in real time through the observation window.

Quick pose measurements of the high-speed and rolling target with different inertia parameters under different Mach wind speeds are conducted. For instance, tests under 0.2 Ma, 0.4 Ma and 1.5 Ma, are completed in the transonic wind tunnel. Afterwards, the pose parameters can be given in the wind tunnel coordinate system, in which the *X*-axis and *Y*-axis are parallel to the airflow direction and the direction of gravity, respectively. Additionally, one point on the airplane cabin door is selected as the origin of the wind tunnel coordinate system and the *Z*-axis is established according to the right-handed system. The measured results for certain types of targets with 1.5 Mach speed are depicted in the following diagrams. The results of the position measurement are shown in Figure 15, and the results of the attitude measurement are shown in Figure 16.

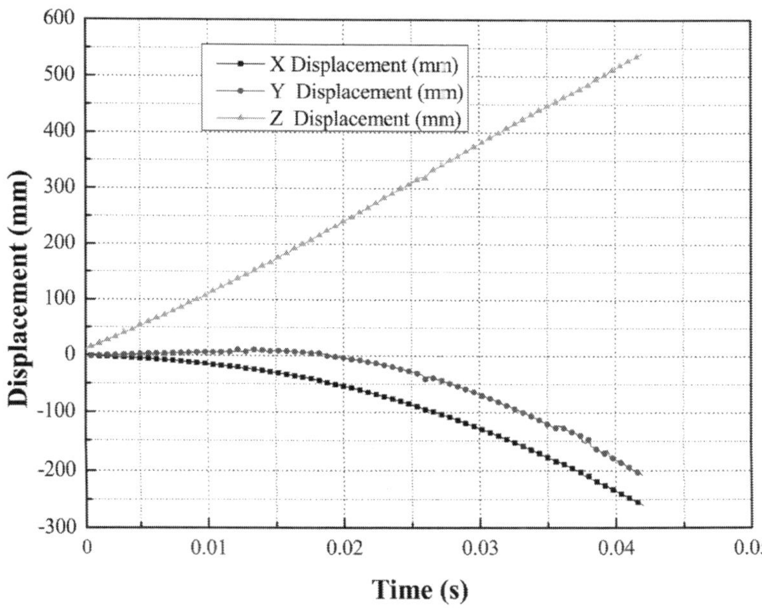

Figure 15. The results of the position measurement in the wind tunnel.

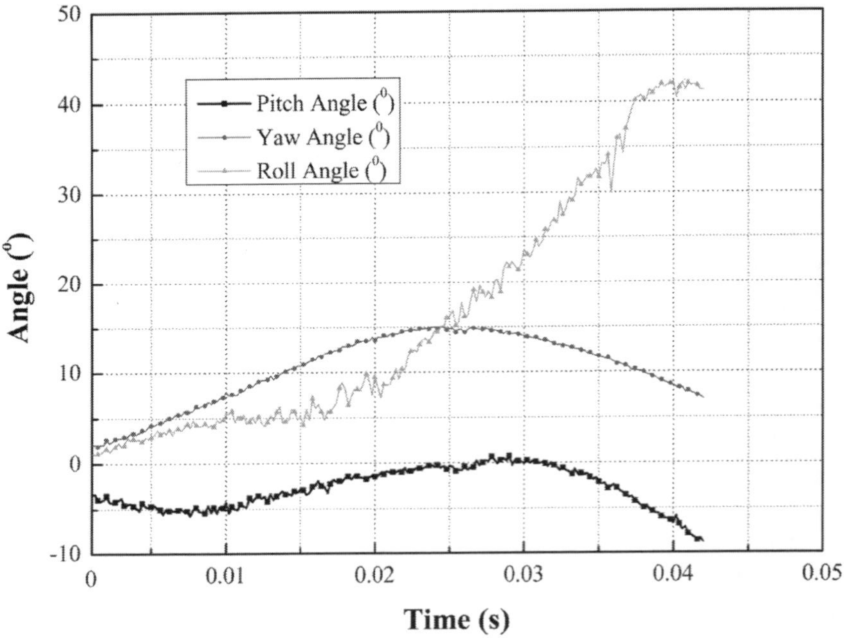

Figure 16. The results of the attitude measurement in wind tunnel.

Experimental results indicate that the quick intellectualized pose measurement of the target can be guaranteed by the proposed method and established measurement system. According to the experimental analysis, three main factors affect the pose measurement accuracy: (a) Problems such as large measuring range, limitation of the resolution of the ultra-high speed camera directly affect pose measurement precision of the target; (b) The extraction accuracy of the markers is affected due to factors such as small volume, surface profile of markers and random deformation during the rolling process.

5.3. Precision Verification Test in Laboratory

In this paper, targets are fixed on a high precision motion control stage. The *X*, *Y*, *Z* axis position measurement precision is verified through shooting the movement of targets when the motion control stage moves a fixed distance along the *X*, *Y*, *Z*axis direction with high precision. Then the attitude measurement precision of the measurement system is verified with the target rotating around the *X*, *Y*, *Z* axis with a fixed angle.

The experimental facilities are shown as Figure 17. For each camera, a high precision motion control stage combining a displacement platform and a rotary platform is built. With a position precision of 0.001 mm, the motion control stage is used to implement the specified moving distance of targets. Twenty four displacement accuracy verification test runs were conducted, and the target moved 20 mm in each run. Twenty four angle accuracy verification test runs were also conducted, where the target rotated five degrees around the three axes, respectively, in each run. The displacement measurement precision in each run is depicted in Figure 18. The measured angle precision in each run is shown Figure 19.

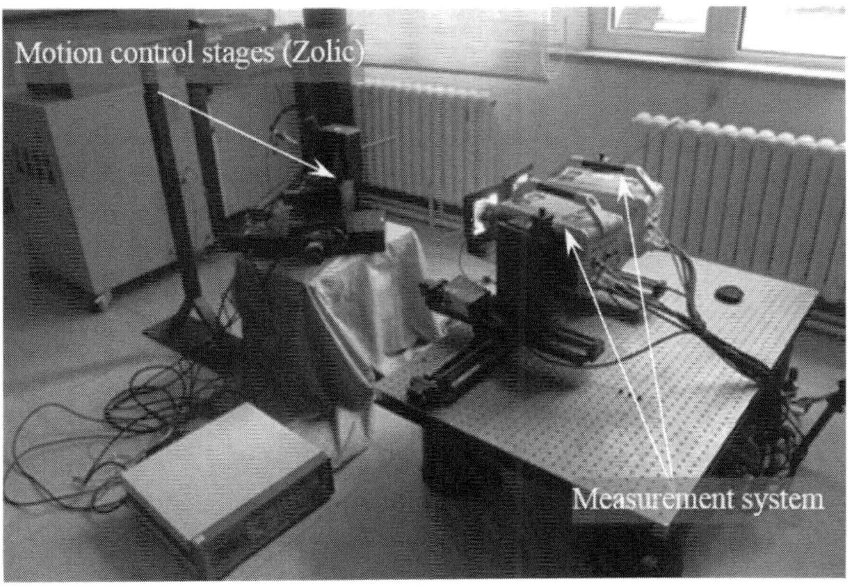

Figure 17. The calibration diagram of measurement system.

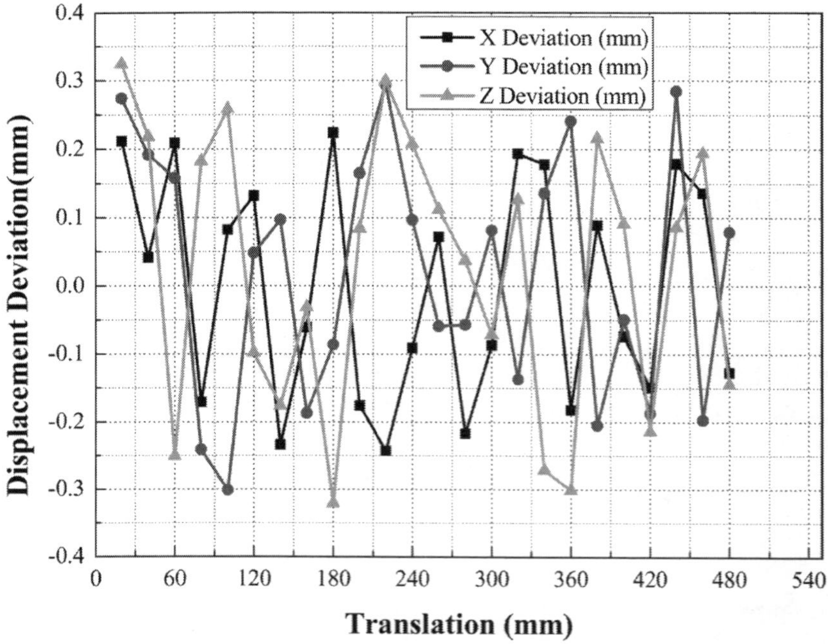

Figure 18. The position measurement precision.

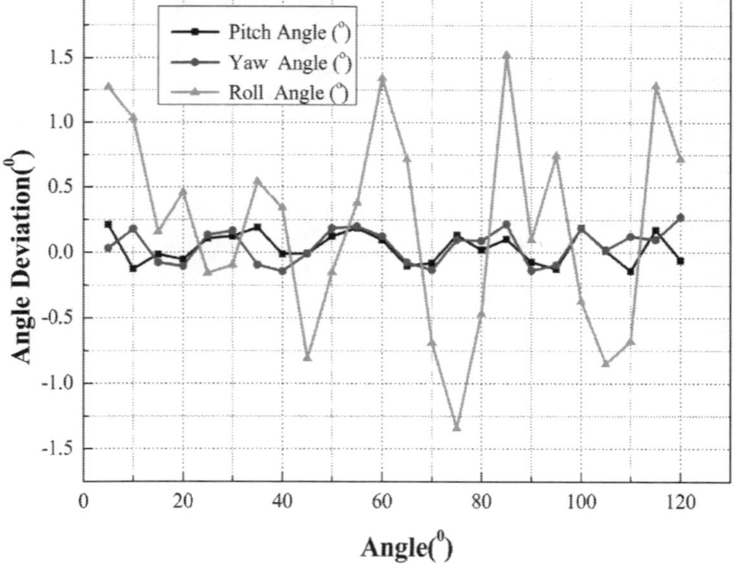

Figure 19. The angle measurement precision.

As shown in Figures 18 and 19, the mean square error of the displacement measurement in the X direction is 0.13 mm, while in the Y direction it is 0.14 mm, and in the Z direction 0.16 mm. The main errors mainly come from errors in the markers' center extraction and camera calibration. The mean square error of the pitching angle precision is 0.115°, and that of the yaw angle precision 0.132°. The high-precision solution of the measurement system for pitching angle and yaw angle results from the high-precision of axis fitting. The roll angle precision is 0.712°, mainly due to the small diameter of the object to be tested. In conclusion, the experimental results indicate that the proposed method can exhibit high accuracy for the position and attitude measurement of high-speed and rolling targets in a wind tunnel.

6. CONCLUSIONS

In this paper, we have presented a pose measurement method for wind tunnel tests to measure high-speed and rolling targets. A high-speed image sequence acquisition method based on ultrathin retro-reflection markers is presented to ensure high brightness and SNR. Ultrathin retro-reflection markers attached on the surface of a target by a rolling-printed method is as thin as 20 μm, guaranteeing the aerodynamic shape of the scaled model.

A novel marker layout on the basis of spatial coding is proposed to ensure that markers are distributed evenly on the target surface and to achieve high-accuracy pose information. This layout overcomes the problem of the information loss of markers when the target is rolling and ensures matching, so that accurate marker recognition can be obtained.

Moreover, a pose parameter computing method is proposed according to the markers' 3D coordinates. Besides, an axis fitting method based on Case Deletion Diagnostic can increase the stability of the measurement system and reduce the errors caused by mismatching.

The experimental results described above indicate that the proposed pose measurement system and method can achieve high accuracy and stability, and meets the requirements of wind tunnel tests. Finally, the rapid pose measurement with high precision for rolling targets is accomplished in a transonic wind tunnel environment, and good results are achieved. Our future research will focus on improving the pose measurement precision by algorithm optimization.

ACKNOWLEDGMENTS

This paper is supported by the National Natural Science Foundation of China (Grant No. 51375075), and the Scientific Research Fund of Liaoning Provincial Education Department (Grant No. L2013035), and the Science Fund for Creative Research Groups of China (Grant No. 51321004).

AUTHOR CONTRIBUTIONS

Zhenyuan Jia and Wei Liu developed the pose measurement method; Xin Ma proposed the ultrathin retro-reflection marker fabrication technique; Wei Liu was responsible for the recognition and matching of images; Xiao Li and Ling Chen participated in the design of the experimental facilities for the wind tunnel; Wenbo Lu, Zhengqu Wang and Xiaochun Cui carried out the wind tunnel tests and the most of the analyses. All authors wrote the paper together.

CONFLICTS OF INTEREST

The authors declare no conflict of interest.

REFERENCES

1. Lee, J.J.; Ho, H.N.; Lee, J.H. A. Vision-based dynamic rotational angle measurement system for large civil structures.*Sensors* **2012**, *12*, 7326–7336.

2. Marrón-Romera, M.; García, J.C.; Sotelo, M.A.; Pizarro, D.; Mazo, M.; Cañas, J.M.; Losada, C. Marcos, Á. Stereo vision tracking of multiple objects in complex indoor environments. *Sensors* **2010**, *10*, 8865–8887.

3. Newman, B.; Yu, S. *Development of a High Accuracy Angular Measurement System for Langley Research Center Hypersonic Wind Tunnel Facilities*; Langley Research Center: Hampton, VA, USA; August; 2003.

4. Park, H.S.; Kim, J.Y.; Kim, J.G.; Choi, S.W.; Kim, Y. A new position measurement system using a motion-capture camera for wind tunnel tests. *Sensors* **2013**, *13*, 12329–12344.

5. Watzlavick, R.L.; Crowder, J.P.; Wright, F.L. Comparison of model attitude systems—Active target photogrammetry, precision accelerometer, and laser interferometer. Proceedings of 19th Advanced Measurement and Ground Testing Conference, New Orleans, LA, USA, 17–20 June 1996.

6. Rock, S.M.; Frew, E.W.; Jones, H.; LeMaster, E.A.; Woodley, B.R. Combined CDGPS and vision-based control of a small autonomous helicopter. Proceedings of the American Control Conference, Philadelphia, PA, USA, 21–26 June 1998; pp. 694–698.

7. Jones, T.; Lunsford, C. Design and development of a real-time model attitude measurement system for hypersonic facilities. Proceedings of 43rd AIAA Aerospace Sciences Meeting and Exhibit, Reno, NV, USA, 10–13 January 2005.

8. Crawford, B.L.; Rhode, M.N. Miniature on-board angle of attack measurement system for hypersonic facilities. Proceedings of 25th AIAA Aerodynamic Measurement Technology and Ground Testing Conference, San Francisco, CA, USA, 5–8 June 2006.

9. Burner, A.W.; Fleming, G.A.; Hoppe, J.C. Comparison of three optical methods for measuring mode deformation. Proceedings of 38th Aerospace Sciences Meeting & Exhibit, Reno, NV, USA, 10–13 January 2000.

10. Graves, S.S.; Burner, A.W. Development of an intelligent videogrammetric wind tunnel measurement system. Processing of Optical Diagnostics for Fluids, Solids, and Combustion, San Diego, CA, USA, 29 July 2001.

11. Murray, E.N.; Jansen, J.B.; Gui, L.C.; Seiner, M.J.; Birkbeck, R. Measurements of store separation dynamics. Proceedings of 47th AIAA Aerospace Sciences Meeting including The New Horizons Forum and Aerospace Exposition, Orlando, FL, USA, 5–8 January 2009.

12. Chen, J.C.; Guo, M.; Ding, Z.L.; Yuan, F. Stereo vision based attitude measurement of aircraft model in wind tunnel. *J. Appl. Opt.* **2010**, *31*, 437–441.

13. Fan, Q.; Wu, Y.; Hui, J. Integrated navigation fusion strategy of INS/UWB for indoor carrier attitude angle and position synchronous tracking. *Sci. World J.* **2014**, *13*, 1–13.

14. Zhang, Z.Y. Camera calibration with one-dimensional objects. *IEEE Trans. Pattern Anal. Mach. Intell.* **2004**, *26*, 892–899.

15. Brown, L.G. A survey of image registration techniques. *ACM Comput. Surv.* **1992**, *24*, 325–376.

16. Christensen, R.; Pearson, L.M.; Johnson, W. Case Deletion Diagnostics for mixed models. *Technometrics* **1992**, *34*, 38–45.

Index